SUSTAINABLE STOCKHOLM

SUSTAINABLE STOCKHOLM provides a historical overview of Stockholm's environmental development, and also discusses a number of cross-disciplinary themes presenting the urban sustainability work behind Stockholm's unique position, and, importantly, the question of how well Stockholm's practices can be exported and transposed to other places and contexts.

Using the case of Stockholm as the pivot of discussions, *Sustainable Stockholm* investigates the core issues of sustainable urban environmental development and planning, in all their entanglements. The book shows how intersecting fields such as urban planning and architecture, traffic planning, land-use regulation, building, waste management, regional development, water management, and infrastructure engineering – together and in combination – have contributed to making Stockholm Europe's "greenest" city.

Jonathan Metzger is Assistant Professor of Urban and Regional Studies at the School of Architecture and the Built Environment at KTH/Royal Institute of Technology in Stockholm, Sweden. He has a background as a practitioner in the urban policy field within both a Swedish and a European context.

Amy Rader Olsson is a researcher at the Division of Urban and Regional Studies, School of Architecture and the Built Environment and with the Center for Sustainable Built Environments at KTH/Royal Institute of Technology in Stockholm, Sweden. She has a background as a consultant in Swedish and international urban policy.

SUSTAINABLE STOCKHOLM

EXPLORING URBAN SUSTAINABILITY
IN EUROPE'S GREENEST CITY

EDITED BY JONATHAN METZGER AND AMY RADER OLSSON

Routledge
Taylor & Francis Group

NEW YORK AND LONDON

First published 2013
by Routledge

Simultaneously published in the USA and Canada
by Routledge
711 Third Avenue, New York, NY 10017

Routledge is an imprint of the Taylor & Francis Group, an informa business

British Library Cataloguing in Publication Data
A catalogue record for this book is available from the British Library

Library of Congress Cataloging-in-Publication Data
Sustainable Stockholm : exploring urban sustainability in Europe's greenest city /
[edited by] Jonathan Metzger, Amy Rader Olsson. -- First edition.
pages cm
1. Sustainable urban development–Sweden–Stockholm. 2. Urban ecology (Sociology)–Sweden–
Stockholm. 3. City planning–Environmental aspects–Sweden–Stockholm. 4. Stockholm
(Sweden)–Environmental conditions. I. Metzger, Jonathan, 1978– II. Olsson, Amy Rader.
HT243.S82S778 2013
307.7609487'3–dc23
2012050193

ISBN13: 978–0–415–62212–7 (hbk)
ISBN13: 978–0–415–62213–4 (pbk)
ISBN13: 978–0–203–76879–2 (ebk)

Typeset in Univers by
Keystroke, Station Road, Codsall, Wolverhampton

DEDICATION

I (Jonathan) wish to dedicate this book to Rafael, Judit, and Tami, who belong to a generation that will hopefully come to find practical ways to truly care for the world. And to Rebecca for her patience with me in finishing this book, and many other things. It is truly more than I have deserved.

I (Amy) dedicate this book to my husband Per and daughters Sofia and Ella, who teach me to meet challenges with curiosity, energy and spirit. I also dedicate this book to the memory of my mentor, John Quigley, for reminding me that "there is no substitute for just doing the work."

CONTENTS

LIST OF ILLUSTRATIONS   xi
LIST OF CONTRIBUTORS   xiii
PREFACE   xvii
ACKNOWLEDGMENTS   xxi

CHAPTER 1   INTRODUCTION: THE GREENEST CITY?   1
Jonathan Metzger and Amy Rader Olsson

1.1 STRUCTURE OF THE BOOK   6

**CHAPTER 2   FROM UGLY DUCKLING TO EUROPE'S
FIRST GREEN CAPITAL: A HISTORICAL
PERSPECTIVE ON THE DEVELOPMENT OF
STOCKHOLM'S URBAN ENVIRONMENT   10**
Björn Hårsman and Bo Wijkmark

2.1 INTRODUCTION   10

2.2 STOCKHOLM IN THE GLOBAL URBAN SYSTEM   11

2.3 STOCKHOLM 1850–1914   14

2.4 A PEACEFUL ERA OF SOWING SEEDS, 1914–1945   23

2.5 REAPING HARVESTS AFTER 1945   29

2.6 THE PAST 40 YEARS   38

2.7 CONCLUSIONS: LOOKING BACK AND LOOKING FORWARD   46

**CHAPTER 3   USING THE CONCEPT OF SUSTAINABILITY:
INTERPRETATIONS IN ACADEMIA, POLICY,
AND PLANNING   51**
*Ulrika Gunnarsson-Östling, Karin Edvardsson Björnberg,
and Göran Finnveden*

3.1 INTRODUCTION   51

3.2 THE CONCEPT OF SUSTAINABLE DEVELOPMENT   53

3.3 THE POLITICAL DISCOURSE ON SUSTAINABLE DEVELOPMENT   58

3.4   WHAT HAPPENS IN REALITY?   62

3.5   DISCUSSION AND CONCLUSIONS   65

**CHAPTER 4   A SUSTAINABLE URBAN FABRIC:
                 THE DEVELOPMENT AND APPLICATION OF
                 ANALYTICAL URBAN DESIGN THEORY   71**
                 Lars Marcus, Berit Balfors, and Tigran Haas

4.1   INTRODUCTION   71

4.2   CONTEMPORARY TRENDS IN URBAN DESIGN   73

4.3   ANALYTICAL THEORIES OF SUSTAINABLE CITIES   82

4.4   TOWARD AN ANALYTICALLY SUPPORTED URBAN DESIGN THEORY   88

4.5   ALBANO: TOWARD PRINCIPLES OF SOCIAL-ECOLOGICAL
      URBAN DESIGN   93

4.6   DISCUSSION: DESIGN THEORY AS AN EPISTEMOLOGICAL
      FRAMEWORK   97

**CHAPTER 5   SUSTAINABLE URBAN FLOWS AND
                 NETWORKS: THEORETICAL AND PRACTICAL
                 ASPECTS OF INFRASTRUCTURE
                 DEVELOPMENT AND PLANNING   102**
                 Folke Snickars, Lars-Göran Mattsson, and Bo Olofsson

5.1   INTRODUCTION   102

5.2   WHAT IS INFRASTRUCTURE?   105

5.3   CHALLENGES IN STOCKHOLM'S INFRASTRUCTURE SYSTEMS   111

5.4   SUSTAINABLE INFRASTRUCTURE SCENARIOS FOR STOCKHOLM   120

5.5   CONCLUDING DISCUSSION   124

CHAPTER 6   THE ECONOMICS OF GREEN BUILDINGS   129
Hans Lind, Magnus Bonde, and Agnieszka Zalejska-Jonsson

6.1   INTRODUCTION   129

6.2   ENVIRONMENTAL RATING TOOLS IN SWEDEN   130

6.3   COMMERCIAL AND RESIDENTIAL GREEN BUILDINGS IN
      STOCKHOLM   131

6.4   ECONOMIC STUDIES OF GREEN BUILDINGS IN SWEDEN:
      NEW CONSTRUCTION   136

6.5   GREENING THE EXISTING BUILDING STOCK   140

6.6   CONCLUDING DISCUSSION   143

**CHAPTER 7   PERFORMING SUSTAINABILITY:
              INSTITUTIONS, INERTIA, AND THE
              PRACTICES OF EVERYDAY LIFE   147**
Ebba Högström, Josefin Wangel, and Greger Henriksson

7.1   INTRODUCTION   147

7.2   INSTITUTIONS, PATH DEPENDENCY, AND DISCOURSES   149

7.3   FROM TEACHER TO PREACHER: THE STORY OF GLASHUSETT   152

7.4   IMPOSING RESPONSIBLE LIFESTYLES IN THE ROYAL SEAPORT:
      ECO-FASCISM OR INDIVIDUAL CHOICE?   156

7.5   THE (IN)VISIBLE WHIP: THE INTRODUCTION AND PERMANENT
      ESTABLISHMENT OF CONGESTION CHARGING   160

7.6   CONCLUDING DISCUSSION   163

**CHAPTER 8   FROM ECO-MODERNIZING TO POLITICAL
              ECOLOGIZING: FUTURE CHALLENGES FOR
              THE GREEN CAPITAL   168**
Karin Bradley, Anna Hult, and Göran Cars

8.1   INTRODUCTION   168

8.2   A GROWING BUT STRAINED STOCKHOLM   170

8.3   FRAMING THE GREEN CAPITAL   171

8.4   FROM LIGHT GREEN SUSTAINABILITY TO POLITICAL ECOLOGY   174

8.5   CHALLENGES FOR A MORE ENVIRONMENTALLY JUST
      DEVELOPMENT   176

8.6   CONCLUDING DISCUSSION: BEING AT THE FOREFRONT
      INTERNATIONALLY   188

**CHAPTER 9   URBAN SUSTAINABLE DEVELOPMENT
               THE STOCKHOLM WAY   195**
               Amy Rader Olsson and Jonathan Metzger

9.1   INTRODUCTION   195

9.2   CREDIT WHERE CREDIT IS DUE: HOW DID STOCKHOLM GET HERE?   197

9.3   PARTICULARITIES AND BLIND SPOTS   199

9.4   KEY STRATEGIC PARAMETERS FOR URBAN SUSTAINABLE
      DEVELOPMENT: LESSONS FROM STOCKHOLM   206

9.5   FINAL WORDS   211

INDEX   212

ILLUSTRATIONS

FIGURES

2.1 Drawing from 1886 of Stockholm's central railway artery passing on a drawbridge across "the Sluice" (Slussen) and further across to the old city   17

2.2 The waterworks at Eriksdal on Södermalm in central Stockholm, 1891   21

2.3 Poster produced by the City of Stockholm in 1946 to deter prospective immigrants from the countryside   27

2.4 The square in Vällingby, internationally renowned "model suburb" built according to the ideal of the so-called ABC city   31

2.5 The redevelopment of the central business district began in the 1950s with the construction of a road tunnel and three intersecting main subway lines   32

2.6 Aerial photograph of Rinkeby at Järvafältet, one of the major developments of the Million Homes Program   34

2.7 The public art in Stockholm's subway system is world renowned, such as Ulrik Samuelson's sculptural installations in the Kungsträdgården station   37

2.8 The 1971 "battle of the Elms" protest against a planned subway entrance in Kungsträdgården in Stockholm's CBD   38

4.1 The SymbioCity concept, an integrated systems approach to sustainable development that is the basis for the Hammarby Sjöstad project   79

4.2 Output from MatrixGreen, a software tool for assessing landscape connectivity   86

4.3 Image created using the Place Syntax Tool, a GIS-based software tool for calculating typical distance measures   92

5.1 Stylized picture of the Stockholm region and its development directions   104

5.2 Fast and slow urban processes at micro and macro scales   109

5.3 Closed-loop system for ground heat extraction and heating/cooling management at Arlanda Airport   114

5.4 Public transport's share of total personal travel in the Stockholm region, 1973–2009   117

5.5 Map of forecast land-use pattern, 2006, if the subway system had not been built, compared to the actual land-use pattern   120

5.6 The interaction between urban structure, user behavior, and welfare development   122

5.7 Four Stockholm region scenarios in the nexus between integration and isolation   123

6.1	Pennfäktaren 11, a commercial building in the Stockholm central business district	132
6.2	The Blå Jungfrun ("Blue Maiden") housing estate in Farsta	134
7.1	Poster from GlashusEtt, the environmental information centre in Hammarby Sjöstad	153
7.2	Perspective sketch of the envisioned environment at the Royal Seaport Beach Park	157
7.3	One of the access roads leading to the central parts of Stockholm, including the information and registration devices	160
8.1	Excerpt from Stockholm's *Vision 2030* document	173
8.2	Emission of greenhouse gases – direct and indirect emissions in tons per capita, 2004	182
8.3	Rough sketch of societal organization and spatial form	184
8.4	The SymbioCity promotion graph "The Swedish Experience," illustrating the phenomenon of a "decoupled" economy	189

TABLES

5.1	Urban infrastructure as technical and organizational artifacts	106
5.2	Properties distinguishing infrastructure capital from other forms of capital	106
5.3	Some components of infrastructure-led scenarios for the Stockholm region	111
8.1	Comparison of the national reporting of CO_2 emissions per capita calculated from a production perspective compared with a consumption perspective	178

BOXES

4.1	Shows photograph of Hammarby Sjöstad in Stockholm, a high-class ecological city district planned for a population of more than 25,000 residents	77
4.2	Shows photograph of the Stockholm Royal Seaport, the city's new major eco-district development area	80
4.3	Shows conceptual sketches for Albano Resilient Campus	95

CONTRIBUTORS

THE MAJORITY OF THE CONTRIBUTORS are active researchers at KTH, the Royal Institute of Technology in Stockholm, Sweden.

Berit Balfors is Professor of Environmental Impact Analysis and Head of the Department of Land and Water Resources Engineering, KTH. Her research is particularly directed toward the application of environmental assessments in planning and decision making, with special focus on biodiversity and ecological assessment.

Magnus Bonde is a PhD student at KTH Division of Building and Real Estate Economics. He holds a civil engineering degree from KTH and is currently working with a project on green commercial buildings, including studies of valuation and leases.

Karin Bradley is Assistant Professor in Urban and Regional Studies at KTH. Her research concerns sociocultural perspectives on sustainable urban development, sustainable lifestyles, environmental justice, and social movements. Her PhD dissertation was entitled "Just environments: Politicising sustainable urban development" (2009) and her postdoc dealt with utopian thought in planning for sustainable futures, resulting in the edited book *Green Utopianism: Practices and Perspectives* (Routledge, forthcoming).

Göran Cars is Professor of Regional Planning at KTH. His professional interests are focused on urban governance, particularly the conditions for planning, decision making, and implementation of urban and regional development projects. A special interest is in negotiations as a tool for collaboration between public and private actors.

Karin Edvardsson Björnberg is Assistant Professor of Environmental Philosophy at KTH. She has a PhD in Philosophy from 2008. Her primary research interests lie in finding out how environmental policies, particularly climate policies, can be made efficient, just, and legitimate. She is currently working on the research program Mistra Biotech – Biotechnology for Sustainable and Competitive Agriculture and Food Systems.

Göran Finnveden is Professor in Environmental Strategic Analysis and Vice-President for Sustainable Development at KTH. He has published more than 60 papers in scientific journals on development and the use of environmental systems analysis tools and environmental policy.

Ulrika Gunnarsson-Östling holds a PhD in Planning and Decision Analysis and is a Researcher at the Division of Environmental Strategies Research at KTH. Her research interests lie mainly in the intersection of planning and futures studies with the perspectives of gender and environmental justice.

Tigran Haas is Associate Professor of Urban Planning and Design at KTH and the Director of the Civitas Athenaeum Laboratory (CAL) applied social research platform. He has a background in architecture, urban design and planning, environmental science, and regional planning. His latest publication is *Sustainable Urbanism and Beyond: Rethinking Cities for the Future*, Rizzoli: New York (2012, editor).

Björn Hårsman is Professor Emeritus in Regional Economic Planning and Head of the Department of Industrial Economics and Management at KTH. He chairs the board of the KTH Centre of Excellence for Science and Innovation Studies and is a board member at the Institute for Management of Innovation and Technology. He earlier served as Dean of the KTH School of Architecture and the Built Environment. His current research is focused on entrepreneurship and location among artists.

Greger Henriksson is a Researcher at the Division of Environmental Strategies Research at KTH. He holds a PhD in Ethnology from Lund University on the sustainability of travel habits. His current research primarily focuses on travel behavior, waste, and ICT as a technological support for more sustainable everyday decision making.

Ebba Högström is a Researcher at Urban and Regional Studies and a lecturer at the School of Architecture, KTH. Her work is on socio-materiality, practices and discourses in architecture and the built environment. Her 2012 doctoral thesis "Kaleidoscopic spaces: Discourse, materiality and practice in decentralized mental health care," focuses on her special interest in institutions and their transformations over time.

Anna Hult is a PhD candidate in Urban Planning at KTH. Her research critically examines the Swedish export of the sustainable city to China. Anna is also co-founder of the Amsterdam and Stockholm-based organization CITIES, where she works on urban research and communication.

Hans Lind is Professor in Real Estate Economics at KTH, with a PhD from the Department of Economics at Stockholm University. Has been project leader for several projects on the economics of green buildings and has also written about valuation methods for green buildings.

Lars Marcus is Professor in Urban Design at KTH School of Architecture and Director of the research group Spatial Analysis and Design (SAD) in the field of Spatial Morphology. He has been one of the co-developers and former chairs of the multidisciplinary international two-year Master's program Sustainable Urban Planning and Design (SUPD) at KTH, and is a founding partner of the consultancy firm Spacescape.

Lars-Göran Mattsson is Professor of Transport Systems Analysis at KTH. He received his PhD in optimization from KTH in 1987. His research includes travel demand modeling, land use and transport interaction, infrastructure and regional development, transport system reliability and vulnerability, road pricing, future studies, and applied systems analysis.

Jonathan Metzger is Assistant Professor of Urban and Regional Studies at KTH. He has a broad social scientific background and concrete experiences from working as a planning practitioner on the regional and transnational levels. His research interests include spatial theory, the ethnography of planning practice, and ecological perspectives on urban planning and regional development.

Bo Olofsson is Professor in Environmental Geology with a specialization in geology and hydrogeology/applied geophysics at the Department of Land and Water Resources Engineering at KTH. His current research interests include groundwater supply in coastal areas and the salinization of groundwater in Sweden.

Amy Rader Olsson is a Researcher in the Division of Urban and Regional Studies at KTH and holds degrees from Princeton University (BA History), University of California Berkeley (Master's degree in Public Policy) and KTH (PhD in infrastructure and planning). Her research focuses on planning institutions for cooperation, conflict resolution, and public engagement. She works with the Center for Sustainable Built Environments at KTH and serves on the Architecture Advisory Council for the Swedish Transport Administration.

Folke Snickars, Professor Emeritus in regional planning at KTH, has published widely on the theoretical and methodological aspects of infrastructure economics, regional systems analysis, and regional planning. He has served as President of the European Regional Science Association, as Editor-in-Chief of *Papers in Regional Science*, and as Coordinating Editor of the book series Advances in Spatial Science (Springer Verlag). He chairs the Scientific Council of the Stockholm Regional Planning Office and is Swedish Chairman of the European Spatial Planning Observation Network.

Josefin Wangel is a Researcher at the Division of Environmental Strategic Research (FMS) and the Centre for Sustainable Communications (CESC) at KTH. She holds a PhD in Planning and Decision Analysis and an MSc in Environmental Science, and has several years of working experience in sustainable urban development. Key fields of research include futures studies for sustainability, socio-material perspectives on energy systems and consumption practices, and policy, planning, and design.

Bo Wijkmark is now retired from more than 30 years of service in leading positions in development planning in the city and region of Stockholm, including as Deputy CEO of the Stockholm City Joint Planning Commission and CEO of Stockholm County Council Office for Regional Planning and Metropolitan Transport.

Agnieszka Zalejska-Jonsson is a PhD candidate in KTH's Division of Building and Real Estate Economics. She holds a Master's degree in Economics and Business Administration and a certification from the Passive House Institute in Darmstadt. Her PhD project focuses on energy-efficient residential buildings: cost, profit, and occupant satisfaction.

PREFACE

IN SEPTEMBER 2010, about 60 researchers from the School of Architecture and the Built Environment at KTH, the Royal Institute of Technology in Stockholm, Sweden's largest and perhaps most influential research environment on urban development, gathered to explore new ways to support urban sustainability research and practice. Several recent reviews of research programs on sustainable development had recognized the challenges of integrating research across disciplinary boundaries and creating platforms for dialogue with practitioners. The academic incentive structure forces urban scientists to prioritize academic, peer-reviewed publications, often representing a single discipline or sector. Students and practitioners repeatedly complain that these are written in academic jargon and do not reflect an understanding of current practice, and that they fail to integrate issues of urban sustainability into some form of broader picture of how challenges and proposed solutions may (or may not) be connected or orchestrated.

At the September workshop, we therefore decided to initiate a multidisciplinary (and later also transdisciplinary) cooperative experiment to produce an integrated account of some of the key challenges regarding urban sustainability, and to do so on the basis of the one case for which we collectively hold the greatest amount of expertise: our own home ground of Stockholm. But the simple convenience of proximity and acquaintance was by no means the deciding factor for focusing upon Stockholm. Rather, our decision to devote an in-depth examination of issues of urban sustainable development to this specific city and region stemmed from the remarkable increase in interest in Stockholm during the past decade or so from an ever-widening circle of international professionals and experts in the area of urban sustainable development. Stockholm had begun to assume a position as a global "good practice" example with regard to sustainable development, and was used as an inspiration and benchmark by cities all over the globe, a position further enhanced by the city's nomination as the first ever winner of the European Green Capital Award in 2010.

Nevertheless, from contacts with students and public officials all over the world we had also learned that far from all of those who took a keen interest in the increasingly famous sustainability measures of Stockholm had the opportunity to personally visit the city to learn first-hand about what is going on there. Even for those of us working and living there, there existed no integrated academic account of Stockholm's development in the area of sustainable urban development that at the same time was accessible to a broad audience. From this insight emerged the idea to produce such an account, which could also function as an initiation for the reader to the entangled complexities of the realm of sustainable urban development – an idea that resulted in this book.

The first challenge was how to organize the complexities of sustainability issues into manageable chapters partitioned in a way that made sense in relation to the broader context of the Stockholm case. There are many ways of conceptualizing sustainable development; one of the best known is probably the "triple perspective" model of sustainable development as comprising economic, social, and environmental sustainability – often translated into a broadly diffused Venn diagram, where the intersection of these categories is supposed to represent the zone where development is multidimensionally sustainable. This is but one of innumerable heuristic devices utilized to concretize the content of the vague concept of sustainable development. For instance, sometimes the extra "dimension" of culture is added. Another way of making the concept of sustainable development touch ground is to approach various sectors of society individually to examine how particular functions in society contribute – or can contribute – to a more sustainable development. Hence, many sustainability books use a sector-oriented approach, examining urban transport systems, waste management, water engineering, and institutional arrangements such as economic incentives.

Yet even though they function as helpful devices for organizing thinking concerning sustainable development, both the triple perspective and the sectoral approach inevitably conceptualize sustainability as the intersection of a number of discrete and self-contained spheres, technical systems, or sectors. Partially resulting from frustration with these established but limiting ways of conceptualizing sustainable urban development, in structuring this book we experimented with finding new ways of relating and developing knowledge across entrenched disciplinary boundaries, to bring together scholars from a broad range of relevant disciplines in their work toward understanding and working for urban sustainable development. Therefore, we have made a particular effort to team up authors from different disciplines so as to integrate knowledge from different fields of academia and practice within each chapter.

Using the case of Stockholm, we explore the core issues of sustainable urban development and planning, in all their entanglements. Intersecting fields such as urban planning and architecture, traffic planning, land-use regulation, building, waste management, regional development, water management, infrastructure engineering – together and in combination – have contributed to making Stockholm Europe's "greenest" city. In addition to a historical overview of Stockholm's environmental development, *Sustainable Stockholm* provides a number of cross-disciplinary thematic chapters presenting the urban sustainability work behind Stockholm's unique position, followed by a chapter discussing how well Stockholm's practices can be exported and transposed to other places and contexts.

The book has been written for an international audience of academics and practitioners with a general interest in sustainable urban development and/or a specific interest in Swedish responses to the challenge of urban sustainability. Although the book is not written as teaching material per se, it could be used as course literature at all levels of study, being perhaps best suited to graduate and postgraduate courses or executive and professional courses in a broad range of disciplines and subjects such as urban and spatial planning, environmental studies, urban studies, regional development, architecture, and geography.

ACKNOWLEDGMENTS

A GREAT MANY INDIVIDUALS deserve our sincere thanks for their contribution to the production of this book. The process of writing multi-disciplinary chapters that include both academic and professional perspectives meant that many more people provided critical input than are credited as authors.

First, we would like to thank the School of Architecture and the Built Environment at KTH. The school created the Research Cluster for Urban Sustainability (now the Center for Sustainable Built Enviornments) that supported, framed, and provided financial support to this project. In particular, the steering committee for the cluster – Göran Cars, Björn Hårsman, Lars Marcus, Lars-Göran Mattsson, Katja Grillner, and Berit Brokking Balfors – helped determine the book's focus, content, and contributors.

Björn Hårsman was a constant inspiration whose experience, expertise, and energy we turned to often, and we were never denied it. Björn's insistence that we listen carefully to the suggestions and input of urban professionals such as Gunnar Söderholm, Head of the Environment and Health Administration of the City of Stockholm, proved invaluable.

We would also like to take this opportunity to thank our anonymous academic colleagues who under quite time-pressed conditions provided critical reading of these chapters and excellent suggestions for improvements. You know who you are – and we extend our sincere gratitude!

We would have been hopelessly overwhelmed by the task of organizing, proofing, and designing this book without the support of Fredrick Brantley at Routledge/NYC as well as our graphic designer Chris Knox and able proofreader Helen Runting. Helena Kyllingstad provided brilliant research support for the Introduction.

Last but most certainly not least, on behalf of all contributing authors we thank the Swedish Research Council Formas for its financial support of this book project (grant no. 2011-1760), as well as previous support for many of the research projects referenced herein.

We look forward to working with all of you again!

Jonathan Metzger and Amy Rader Olsson

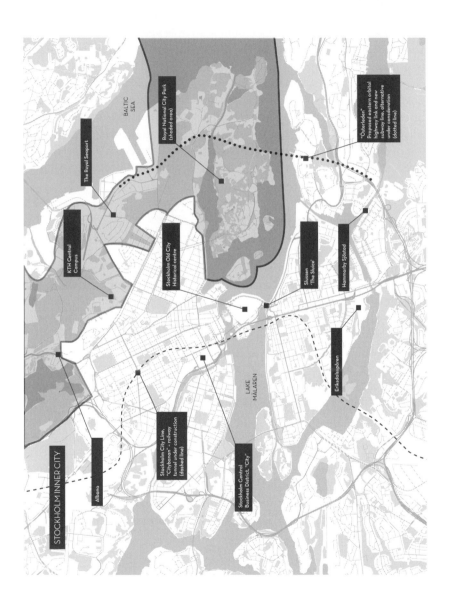

STOCKHOLM INNER CITY

Albano

Stockholm City Line,
"Citybanan", railway
tunnel under construction
(dashed line)

Stockholm Central
Business District, "City"

KTH Central
Campus

The Royal Seaport

Stockholm Old City
Historical centre

Royal National City Park
(shaded area)

Slussen
"The Sluice"

Hammarby Sjöstad

Eriksdalspåren

"Österleden"
Proposed eastern orbital
highway link and new
subway line, alternative
under consideration
(dotted line)

BALTIC
SEA

LAKE
MÄLAREN

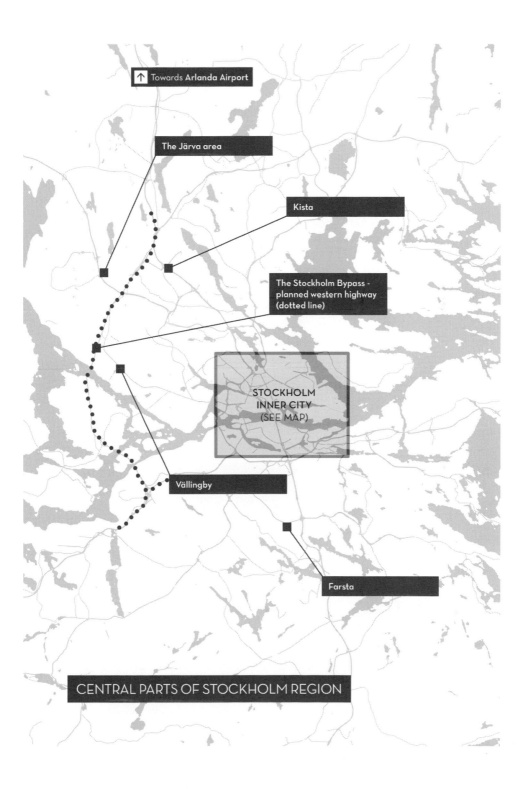

↑ Towards **Arlanda Airport**

The Järva area

Kista

The Stockholm Bypass - planned western highway (dotted line)

STOCKHOLM
INNER CITY
(SEE MAP)

Vällingby

Farsta

CENTRAL PARTS OF STOCKHOLM REGION

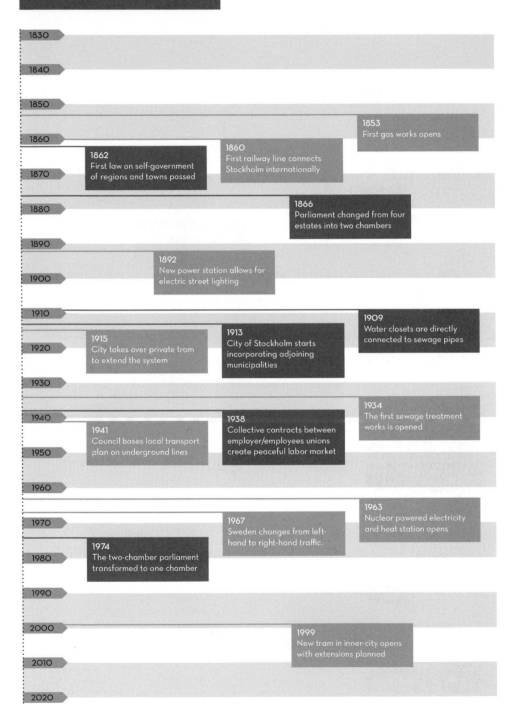

STOCKHOLM TIMELINE

1830

1840

1850

1860
1853
First gas works opens

1862
First law on self-government
of regions and towns passed

1860
First railway line connects
Stockholm internationally

1870

1880
1866
Parliament changed from four
estates into two chambers

1890
1892
New power station allows for
electric street lighting

1900

1910
1909
Water closets are directly
connected to sewage pipes

1920
1915
City takes over private tram
to extend the system

1913
City of Stockholm starts
incorporating adjoining
municipalities

1930

1940
1934
The first sewage treatment
works is opened

1941
Council bases local transport
plan on underground lines

1938
Collective contracts between
employer/employees unions
create peaceful labor market

1950

1960

1970
1963
Nuclear powered electricity
and heat station opens

1967
Sweden changes from left-
hand to right-hand traffic.

1974
The two-chamber parliament
transformed to one chamber

1980

1990

2000
1999
New tram in inner-city opens
with extensions planned

2010

2020

1830

1840

1842
Compulsory school education
for all children

1850

1866
"Lindhagen-plan" for the
extension of the city formally
repelled but implemented

1859
City takes responsibility for
street cleaning and refuse

1861
First water works opens

1860

1870

1874
New Health Code, Building
Code and Fire Code created

1877
First private-owned tramway
line opens

1880

1890

1904
City buys land in adjoining
districts to extend the city

1900

1910

1912
Stockholm is Europe's
telephone densest city

1920

1920
Administrative reforms make
the ruling of Stockholm more
effective and democratic

1921
Universal suffrage to
Parliament and councils

1930

1952
Radical general plan for the
reconstruction and expansion
of the city formally repelled
but largely implemented

1940

1950

1952
2500 Swedish towns and
municipalities amalgamate to
less than 1000

1960

1966
European motorways E4/E20
through Stockholm open

1971
County of Stockholm takes
over responsibility for healthcare, transport and planning

1970

1980

1987
New national planning
and building act

1990

2000

2006
Trial of congestion charging
to reduce traffic takes place

2010

2010
New tram line from the centre
to National City Park opens

2020

FACTS ABOUT STOCKHOLM

POPULATION

As of 2010, the Stockholm metropolitan area was home to approximately 22% of Sweden's population

Year	Stockholm City	County of Stockholm
1980	647 214	1 528 200
1990	674 452	1 641 669
2000	750 348	1 823 210
2009	829 417	2 019 182
2010	847 073	2 054 343

HOUSING

Calculated possible demand of new residences in the region (2010-2030) **319 000**

Planned construction of new residences in the region (2012-2021) **149 842**

% of realization of planned construction of residences 10 years after plan draft (2012) **50%**

ENERGY SUPPLY

Total energy supply for the city of Stockholm 2009 (GWh)

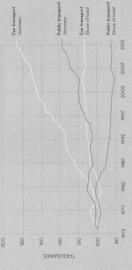

Crude oil 5420 26%

Coal 1935 10%

Biofuels 2912 14%

Electricity produced outside Stockholm 5918 29%

Waste 1074 5%

Heat pump 2125 10%

Other 1128 6%

GREEN SPACE

90% of residents can walk to a public green space within 300m

TALLEST BUILDINGS

Some of the most identifiable buildings on Stockholm's skyline

City Hall 106m

Klara Church 116m

Kaknäs Tower 155m

Kista Science Tower 158m

PUBLIC TRANSPORT

Storstockholms Lokaltrafik (SL) organizes and provides an extensive system of public transport for residents and visitors. It combines bus, tram, light rail, metro and tram services allowing travel within the city and region.

Right
Number of passenger journeys (millions)

1992	2011
539	718

Price (SEK) of one months pass for transport system

1995	2011
355	790

Cars per 1000 residents in the Stockholm region

1980	2011
301	392

Below
Development trends for passenger journeys in region, 1973 (index=100)

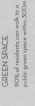

Car transport Journeys

Public transport Journeys

Car transport Share of total

Public transport Share of total

THOUSANDS

200
180
160
140
120
100
80

1972 1977 1982 1987 1992 1997 2002 2007 2012

CHAPTER 1
INTRODUCTION
THE GREENEST CITY?

Jonathan Metzger and Amy Rader Olsson

THE EUROPEAN UNION inaugurated the European Green Capital Award in 2010. The annual prize recognizes the consistent record of high environmental standards and commitment to ongoing and ambitious goals for further environmental improvement of one city within the European Union, which can then function as a role model to inspire and "promote best practice experiences in all other European cities" (European Green Capital, 2009). The first city to receive this award was Stockholm, the capital of Sweden.

Positioned on the Baltic rim, on the northern outskirts of Europe, Stockholm has long been renowned for its distinctive waterfront and extensive foliage, truly making this "Venice of the North" worthy of the moniker *Green* (and blue!) *Capital* in the literal sense of the term. But of course, in relation to urban development the adjective "green" also implies so much more, functioning as a synonym for *sustainability*, or *sustainable development*. The classic definition of sustainable development, as formulated in the Brundtland Report (World Commission on Environment and Development, 1987), states that sustainable development is development that "meets the needs of the present without compromising the ability of future generations to meet their own needs." In many contexts, the term is also used in a broad sense to denote action toward securing, sustaining, and developing the global preconditions for human life across multiple generations, while simultaneously paying attention to human well-being and prosperity, as well as safeguarding sensitive ecosystems and finite natural resources (for a more detailed discussion, see Chapter 3).

Cities and urban areas are increasingly in focus regarding questions of sustainable development. The United Nations estimates that in 2008 more than 50 percent of the world's population lived in cities. This share is steadily rising, and in many parts of the world the pace of urbanization is still rapidly accelerating – especially in the so-called Global South. Over 75 percent of EU

citizens already live in urban areas. Cities themselves are not only distinguishable places that can be easily pinpointed on a world map, but also nodes,
or rather bundles of nodes, within networks that both constitute and are
constituted by innumerable flows of people, ideas, and resources – flows at
the intersections of which the phenomenon of conurbation is generated.
These networked relations sometimes span the globe: what happens in one
city can have effects on other places that may physically be located on the
other side of the earth. Hence, the city can be conceptualized as both a local
phenomenon and a global one that has geographically distributed effects.
What we do in one city – how we build, how we consume, how we act – often
has repercussions globally (in terms of climate change, ozone, etc.) and also
very concretely for other specific, sometimes distant, places.

Historically, cities have generally been associated with adverse
environmental effects and therefore primarily have been seen as a problem
that must be dealt with. Today, the vast bulk of environmentally detrimental
production and consumption is still concentrated in urban areas. However,
increasingly, urbanization and city development are also beginning to be seen
not only as problems but also as important parts of the solution in the quest
towards reaching worldwide sustainable development – for instance, in
relation to issues such as climate change mitigation, energy conservation,
protection of arable land, and water management. This is reflected in the ten
key indicators of sustainable urban development used to determine the
winner of the EU Green Capital Award: *local contribution to global climate
change*, *local transport*, *availability of local public open areas and green areas*,
quality of local ambient air, *noise pollution*, *waste production and management*, *water consumption*, *wastewater treatment*, *environmental management by the local authority*, and *sustainable land use*.

As a factor in motivating its choice to award Stockholm the Green
Capital Award, the jury noted that Stockholm has an "outstanding, long
historical track record of integrated urban management also confirmed by its
ongoing credible green credentials," and, further, that "ambitious plans for
the future clearly demonstrate continuity" (European Green Capital, 2009).
The independent expert panel that made an in-depth evaluation of the
applications for the award chose to highlight a handful of areas in which
Stockholm was judged to excel. To begin with, the panel underscored
overarching policy structures and practices, such as the city council's holistic
vision with ambitious long-term targets and performance indicators, as well
as the solidly budgeted environmental program with its focus on combining
economic growth with an environmental sensibility. A further area lauded
by the panel was Stockholm's integrated city management system, which
includes environmental factors and goals as an integral part of the city
administration's management routines, and makes environmental issues
constantly visible and present in the city's budget, operational planning,
reporting, and monitoring.

After pointing out the history of good results in relation to executed environmental policies, the evaluation panel turned to more concrete and delimited systems and structures, such as the *green structure* of the city, with 95 percent of the population living only 300 meters or less from green areas; its integrated *waste management*, with a high level of recycling, particularly of bio-waste; and its *transport* system, with congestion charging to reduce car use, well-functioning public transportation, and pro-cycling policies. Finally, the panel emphasized an aspect of Stockholm's application that is of a somewhat different nature than the other performance areas: namely, the *communication strategy* of the city in relation to sustainability efforts. The panel was impressed by the level of commitment and eagerness on behalf of the city to "share its experiences and act as an inspiration to other cities" (European Green Capital, 2009).

The selection of Stockholm as the inaugural award winner probably did not surprise many in the international community of urban policy experts and planners, where the association between the terms "Stockholm" and "sustainable development" has been firmly established for decades. Even before receiving the recognition and publicity associated with the award, Stockholm has in professional circles been considered one of the few major metropolitan areas in the world that is on a path toward sustainable development. Stockholm has been recognized for its innovative take on urban sustainability, supported across the political spectrum and combining grand visions and goals (such as becoming 100 percent fossil-fuel-free by 2050) with practical interventions and measures (such as congestion charging and eco-profiled major redevelopment areas).

For sure, not all of Stockholm's sustainability scores are exemplary. It is no surprise that Stockholm does an excellent job of managing its scarce or expensive resources, such as energy for heating homes in its cold climate. However, Stockholm is well behind other European cities in areas where its resources are abundant and effects less costly in the near term. For example, Stockholm produces more municipal waste per capita than many other European cities, and also compares very poorly to other cities with regard to water-use efficiency. Nevertheless, it can be argued that environmental sustainability has become one of Stockholm's defining characteristics and is considered a major source of its attractiveness to residents, tourists, and firms.

Some of the roots of Stockholm's commitment to sustainable urban development can be traced to the social engineering approach of the Swedish modernist urban planning tradition – although this is a somewhat contested legacy, as the high modernist vision of the city also built upon ideas such as the car-based society, the physical separation of urban functions, and other notions that are today considered directly counterproductive to sustainability. Further, it can be speculated that the firm association between sustainable development and Stockholm that exists today might not originally fully stem from the city's own achievements, but might partly also be a residual effect of the city's hosting of an international conference in the summer of 1972, the United Nations Conference on Human Environment, colloquially known simply as "the Stockholm Conference." As a direct predecessor to global environmental gatherings that followed, such as Rio de Janeiro in 1992 or Johannesburg in 2002, the Stockholm Conference was the first ever international top-level gathering that seriously addressed what has since come to be called *sustainable development*. Stockholm was chosen to host the conference primarily in its capacity as the capital city of the nation of Sweden, as a corollary of the Swedish national government's efforts to make the conference happen – and the choice of Stockholm as the venue for the conference was subsequently the result of national Swedish initiatives in the global arena rather than a reflection of any local interest or achievements on behalf of the Stockholm leadership and administration.

The city's reputation as a hotspot for sustainable development was further reinforced by Stockholm's bid for the summer Olympics in 2004. At the advice of marketing consultants, and in part to appease strong NGO opposition to the bid, Stockholm took a strong eco-profile in its candidacy, building upon inspiration from Sydney's successful branding of the summer games of 2000 as a "green Olympics." Among other things, the proposal included a radical update of the existing infill construction plans for an extensive inner-city brownfield waterfront site so as to comprise a cutting-edge, systemically integrated Olympic Village of previously unheard-of environmental standards. Stockholm's Olympic bid eventually failed, and the games were awarded to Athens. But construction of the proposed eco-profiled Olympic village, Hammarby Sjöstad, still went ahead – as well as a new light rail line linking the area to Stockholm's extensive public transit system. The high-sustainability ambitions of the now world-renowned neighborhood neatly aligned with the governing Social Democratic Party's "green welfare state" policy doctrine from the mid-1990s (*Gröna Folkhemmet*, literally "Green People's Home"). The city council also received substantial financial support from the national government to further develop and implement the eco-profile of the area – which was seen as a flagship project for the national environmental policy orientation. Since then, and particularly after the sweeping national and local electoral victory of a conservative-led coalition in 2006, the Hammarby Sjöstad area has also been utilized and developed as a showcase for Swedish clean technology and eco-technology, as part of a national export strategy for such products that has now been in place for roughly a decade.

Successively, consensus has developed across the political spectrum in Stockholm's city council, positing urban sustainable development as a key goal for the city's future. Serious debate still remains concerning what in practice constitutes such development, what standards to aim for, and how best to work towards reaching this broad and somewhat vague goal. Moreover, broad political backing for a sustainable development focus in Stockholm is hardly of a completely altruistic or idealistic nature. Just as the Stockholm Conference in 1972 aimed both at protecting the global environment and at strengthening Sweden's international reputation and geopolitical position, the driving forces behind the "greening" of Stockholm's policies and investments reflect genuine environmental concern but are to some extent also fueled by recognition of the branding advantages and economic opportunities offered by establishing a green reputation. It is therefore not impertinent to ask how the broad political agreement to focus strongly on urban sustainable development can be maintained if future political actors judge there to be more goodwill and economic opportunities to be gained by focusing on some other branding profile for the city. Mistrustful of the political objectives behind Stockholm's urban sustainability push, critical observers have therefore dismissed Stockholm's green profile as somewhat of a marketing ploy. To an extent, it is difficult not to agree that the global narrative presently being woven around Stockholm as a beacon of urban sustainability sometimes may appear a bit uncritical and almost utopian, with stories often carrying a tinge of "in a magical land far, far away in the well-ordered, distant North." These narratives commonly uncritically present Stockholm as a "best practice" example in the field of urban sustainable development, showcasing the city as some form of universal model to learn from – while, in the process, generating invaluable amounts of goodwill and brand value for the city and its firms.

Nevertheless, just because the "green" profile of Stockholm contains a marketing component, this does not necessarily mean that it also lacks substance. On the contrary, it could actually be argued that the Janus face of Stockholm's environmental ambitions actually neatly illustrates the two sides to the concept of sustainable development: to safeguard the living environment, while in the meantime also generating the necessary pre-conditions for human welfare and prosperity. Things may, nevertheless, be a little bit more complicated than this. For, even if Stockholm is recognized as the leading European city with regard to sustainable urban development, in this field "best" might actually be far from good enough if the aim is to achieve truly sustainable development on a global basis.

Of course, even leading academic researchers are divided in their position on the current status of things. Reflecting differences similar to those among the general public, some academics are quite sanguine about the current situation and future prospects, while others see a need for more fundamental changes in the fabric of the city and social life in general to achieve truly sustainable cities. These different perspectives are, to various degrees, also reflected among the researchers who have contributed to this book. Some choose to focus on concrete improvements; others express direct frustration about gaps and conflicts related to current sustainability efforts in Stockholm. There is no consensus among researchers in this question.

Therefore this book highlights a range of perspectives and tones, some more hopeful and others concerned about current development trajectories. It is up to the reader to balance the scales for her- or himself at the end of the story. Taking into consideration both the marketing and storytelling side of the Stockholm sustainability saga and the nitty-gritty work being done to achieve concrete urban transformations, the purpose of this book is therefore to attempt an evenhanded and sober assessment of Stockholm's achievements and shortcomings in the area of urban sustainable development. It also aims to provide a broad introduction to some of the key challenges and issues related to contemporary urban sustainable development, illustrated by relevant and inspiring examples from Stockholm.

Of necessity, a book such as this can only ever scratch the surface of a wide and crucial research topic such as sustainable urban development, despite focusing on only a single city. Our hope is therefore that readers will be inspired to pursue deeper investigations in the particular subareas of sustainable urban development introduced in this book – in Stockholm or elsewhere. Some of the technologies and practices employed in Stockholm's pursuit are most arguably truly unique – or at least context-specific. Nevertheless, we hope that Stockholm's story can inspire students, practitioners, and scholars to consider what can and must be done today, in Stockholm and in cities around the world.

1.1 STRUCTURE OF THE BOOK

As a result of both the multidisciplinary cooperation that has formed the basis of this book and its foundational ambition to get beyond established but highly limiting conceptualizations of sustainable development, we have chosen as our primary focus the interplay of networks and the geographies they generate; such as *regions*, *functional urban areas* and *neighbourhoods*. In the book, we therefore refer to "Stockholm" and "the Stockholm region" both as formally recognized administrative units and also as specific functional geographies of labor and housing markets, watersheds, energy flows, and similar systemic "hanging-togethers." We are interested in tracing and mapping how different systems together generate urban structures and patterns that are concomitantly *technical*, *social*, *political*, *economic*, *cultural*, and *natural* in various aspects and from different perspectives; dimensions of the city which we see as being inevitably entangled.

To pursue this basic relational intuition, the chapters of this book are organized thematically rather than sectorally. The chapters cut across common disciplinary boundaries – incorporating aspects of urban planning, architecture, economics, engineering, environmental studies, and more – to discuss critical issues of urban sustainability. Each of the seven thematic chapters that follow has been authored by a constellation of urban scientists from a wide range of disciplines and departments, and has been reviewed or co-authored by practitioners in several fields.

Chapter overview

Given Stockholm's sustainability achievements, it can be a shock to realize that at the beginning of the modern era, Stockholm was still one of Europe's dirtiest and most unsanitary places. In Chapter 2, "From ugly duckling to Europe's first green capital: A historical perspective on the development of Stockholm's urban environment," we follow the city's fascinating journey from being a decrepit and relatively peripheral city at the outskirts of Europe in 1850 to becoming today's world-renowned "green" metropolis, with an important position in global economic networks. This chapter describes the interplay between social, political, natural, and technical aspects over the past 160 years of Stockholm's historical development toward urban sustainability. The chapter also frames this narrative within contemporary urban economic geography and institutional theory, showing how Stockholm's general development can be understood both as a function of specific local conditions and traditions, and as the result of more general, global trends and development trajectories.

In view of the fact that environmental concerns today assume center stage in Stockholm's urban management and planning, it is fascinating to learn that until just a few decades ago the concept of *environment* did not even exist in the city's political vocabulary. Chapter 3, "Using the concept of sustainability: Interpretations in academia, policy, and planning," gives a broad introduction to the conceptualization of sustainability in different spheres of society, with a particular focus on Stockholm. The chapter discusses the difference between various conceptualizations and interpretations of sustainable development, and how these can affect and direct concrete practices and interventions. Following a discourse-analytical approach, the chapter shows how "the environment" became a top priority on the city's political agenda and delves into the complex and highly political struggles around how the concept of sustainable development should be interpreted in policy and practice.

In 1930, the modernist ideals of providing quality housing and public facilities for families of modest means attracted 4 million visitors to the Stockholm Exhibition. By the 1950s, this ideal had extended to the creation of entirely new neighborhoods that provided necessary local services, housing, and transit links to the inner city. Since then, Stockholm has become known for its signature flagship neighborhood developments. One of the best-known examples is the internationally celebrated "model suburb" Vällingby. This reputation has been further reinforced in recent times through the closed-loop systemic integration of Hammarby Sjöstad, and the holistic sustainability vision of the emerging Royal Seaport. Chapter 4, "A sustainable urban fabric: The development and application of analytical urban design theory," provides an introduction to the most influential contemporary urban design theories affecting and reflected in these developments, with a particular focus on comparing and contrasting variants of sustainable urbanism. Set in a design-theoretical framework arguing for a wider range of knowledge forms, it introduces some of the theoretical concepts and concrete analytical tools that have been developed at KTH (*Kungliga Tekniska Högskolan*, the Royal Institute of Technology) and utilized in Stockholm to facilitate sustainable urban design, including various social-ecological methods and models for analysis and visualization.

Though Stockholm is well known for its eco-neighborhoods, the Green Capital Award chose to highlight the overarching, systemic aspects of Stockholm's sustainability approach. Key urban infrastructure systems such as traffic and telecommunications, water circulation and purification, waste disposal and recycling, and power and heating are perhaps not as visually striking and easy to point out on a map as a new residential area, but are at least as important to sustainable urban development. Chapter 5, "Sustainable urban flows and networks: Theoretical and practical aspects of infrastructure development and planning," examines the sustainable development and management of urban infrastructural networks. Opening with a discussion of the nature and definition of urban infrastructure and the uncertainties involved in planning it, the chapter examines how Stockholm consciously works toward developing its infrastructure in a more sustainable direction.

Chapter 6, "The economics of green buildings," examines another way in which green ambitions are materially manifested by investigating urban sustainable development from the perspective of the individual building. The chapter presents a broad introduction to various building technologies, methods, standards and classification systems for *green building*. The chapter further discusses how green building has gone from being an almost unknown concept among real estate developers and investors in Stockholm to becoming something that is increasingly seen as a necessity in a market-driven process where "green" choices are made because of their profitability rather than by force or conviction.

How sustainable are urban lifestyles and how are these affected by the city's technical systems, built structures, policy interventions, cultural values, and collective life patterns and individual preferences? Chapter 7, "Performing sustainability: Institutions, inertia, and the practices of everyday life," examines the interplay between institutional arrangements and policy interventions, and the everyday life choices and behaviors of city-dwellers. The chapter highlights the concrete effects of policy initiatives, technologies, and instruments by examining how contemporary innovative urban policy interventions aimed at enhancing sustainability, such as congestion charging, actually "touch ground" through the actions of the inhabitants of the city.

While recognizing Stockholm's achievements in the field of urban sustainable development, Chapter 8, "From eco-modernizing to political ecologizing: Future challenges for the green capital," argues that in order to live up to its political ambitions to be a world-leading green city, Stockholm cannot rest on its laurels but rather must up the ante to tackle a range of challenging issues and difficult trade-offs. In a close examination of Stockholm's current overarching sustainability approach, a number of key issues and developments that currently fall outside the frame are highlighted. Arguing from the perspective of political ecology, the chapter concludes that profound, long-term urban sustainability demands much more than new smart technologies and innovative solutions to urban construction and development. Rather, it most probably entails a complete rethinking of our global relationships, both with humans in other places and with the ecosystems that provide the local and global preconditions for life.

The concluding chapter of the volume, Chapter 9, "Urban sustainable development the Stockholm way," builds upon the insights provided in the previous chapters of the book and further extends the analysis of these by reflecting upon what substantial and generalizable lessons can be drawn from Stockholm's specific experiences in its ambitions toward urban sustainable development. After reviewing some of the specific particularities and blind spots of the current loosely sketched "Stockholm take" on sustainable urban development, it concludes with a presentation of conceptual pairings that highlight some of the difficult balancing acts and tensions that must be negotiated in any policy development scheme aiming at promoting urban sustainable development.

REFERENCES

European Green Capital (2009) *The Expert Panel's Evaluation Work and Final Recommendations for the European Green Capital Award of 2010 and 2011.*
World Commission on Environment and Development (1987) *Our Common Future.* Oxford: Oxford University Press.

CHAPTER 2
FROM UGLY DUCKLING TO EUROPE'S FIRST GREEN CAPITAL

A HISTORICAL PERSPECTIVE ON THE DEVELOPMENT OF STOCKHOLM'S URBAN ENVIRONMENT

Björn Hårsman and Bo Wijkmark

> Towns are like electric transformers. They increase tension, accel-
> erate the rhythm of exchange and constantly recharge human life.
> (Fernand Braudel, *Civilization and Capitalism*)

2.1 INTRODUCTION

METROPOLITAN REGIONS AND URBAN GROWTH are often associated with social imbalances, congestion, and pollution. But contemporary urban research has also shown that metropolitan regions harbor the knowledge, creative energy, and entrepreneurial spirit needed to introduce new technologies and jobs. In that respect, they certainly have the potential to decrease the extent of current sustainability gaps.

Stockholm is one such innovation- and knowledge-oriented metropolitan region and was appointed the first "Green Capital of Europe" in 2010. If we look at the Stockholm of around 160 years ago, the difference is astounding: it was one of the poorest cities in Europe and its inhabitants faced appalling living conditions; the expected life span was only 25 years for men and 31 years for women. Few, if any, observers would have characterized Stockholm as a creative and environmentally sound city in 1850.

This chapter describes the main features of economic, social, and political development in Stockholm since 1850. The aim is threefold: first, to provide an overview of the historical transformation, making it easier to understand the other chapters in the book; and second, to discuss and indicate possible reasons behind the transformation. The third, and related, aim is to shed light on path dependencies – that is, the extent to which historical decisions and processes might exert an influence on current decision making.

The chapter is structured in six sections. The first positions Stockholm from the perspective of contemporary theories of economic geography. The following four sections describe the development of Stockholm during the consecutive periods 1850–1914, 1914–1945, 1945–1970, and 1970–2010. Each section tries to identify the most important economic, social, and political changes from a backward-looking as well as a forward-looking perspective. The last section makes use of the historical exposé to provide some speculative remarks on Stockholm's future development in terms of sustainability.

2.2 STOCKHOLM IN THE GLOBAL URBAN SYSTEM

The emerging current understanding of cities is as parts of nested networks, parts that are integral to both large regional landscapes and the global metabolism, and that affect the world in ways that may have non-linear consequences (e.g. Batty, 2008). On the one hand, the consumption and lifestyles of groups with relatively high incomes – most obvious in the urban areas of industrialized countries – contribute disproportionally to anthropogenic impacts on climate and the environment. On the other hand, cities are wonders of opportunity, efficiency, and creative energies, and harbor some of humanity's finest achievements.

Urbanization and shifts to larger urban settlements are processes that are predominantly driven by economic forces. People move to cities in order to find jobs or better educational possibilities, or to experience the larger variety of different kinds of services. Firms establish and expand in cities because of various economy-of-scale phenomena. Transport infrastructure is another structurally important factor: the higher the relative accessibility of a city, the more attractive it tends to be. As was demonstrated by Andersson (1985), domestic migration to Stockholm during the twentieth century has decreased the initial income advantage of living there. However, at the same time successive infrastructure investments have increased Stockholm's accessibility in relation to other Swedish regions. Owing to the increasing accessibility and possibilities to reap the fruits of economies of scale, the Stockholm region is still attracting migrants from other parts of the country.

Closeness and accessibility allow households and firms to *share* indivisible facilities and the gains from a many-sided specialization, to achieve a better *matching* between "buyers" and "sellers" in different markets, and to make use of face-to-face contacts for *learning and innovations* (Henderson and Thisse, 2004). Representing the "New Economic Geography," these theories also provide deeper insights into the interdependencies between land use and accessibility in metropolitan areas. By way of example, a recent study published by the Swedish Ministry of Finance (see Börjesson, 2012) suggests that the population density of the neighborhoods closest to inner Stockholm would have been significantly higher had not the City, as early as in the 1940s, decided to invest in a subway system. Instead of adding new rings of housing around the inner city, new suburbs were established along the new subway lines.

To recognize the benefits of agglomeration is not to imply that rapid urbanization and the increasing number of large urban regions will automatically make humankind better off. City formation imposes costs in the form of congestion, different kinds of pollution, slums, crime, etc., and those costs typically rise disproportionally with city size. One example is the heat island effect. Owing to loss of vegetated areas and emissions of waste heat, cities are significantly warmer than surrounding rural areas. This makes them more vulnerable during warm weather. According to Stone (2012), more than 70,000 people in Europe died during the 2003 summer heat wave. Cities certainly do not cause heat waves, but they amplify them and hence add to the health problems and death toll related to episodes of hot weather.

Additionally, urban growth in itself brings tensions with regard to capacity, for example in the form of shortages in housing and water, and these tensions include inequalities in the distribution of wealth and well-being. Currently there are about a billion slum dwellers living in large and fast-growing cities, mainly in developing countries. Rapid urbanization may also cause social problems in countryside areas experiencing population decline.

The "public costs" exemplified here indicate that cities need mechanisms and institutions that can be used for closing gaps between private and social benefits and costs; and between local, national, and global benefits and costs. Lacking such policy instruments, some cities will probably grow larger than they should, and the pace of urbanization may sometimes be faster than it would have been had the social and environmental effects been properly addressed by governments at different levels, by urban stakeholders, and by urban citizens. Some countries, including Sweden, have implemented various regional policy measures, for example relocating of government agencies to slow down metropolitan growth rates and further growth in smaller cities. However, this kind of policy seems by and large to have had at best marginal effects.

In a global systems perspective, large urban regions distinguish themselves in two respects. First, they have a higher degree of self-sufficiency, which means that interdependencies that are vital for development and sustainability operate in an encircled, "local" territory, with – in principle – greater opportunities to design governance systems aiming at improved resource efficiency and reduced social and environmental tensions. Second, large urban regions are characterized by more wealth, and thereby they initiate larger than average long-distance flows per capita. A larger share of local interdependencies in urban regions provides an opportunity to reshape and design the entire social fabric in the urban context. The larger share of global interurban long-haul transport constitutes an emission problem that is urban-specific. However, it also offers an opportunity, because transport flows between large urban regions can make use of technical solutions that only apply to large-scale flows. The introduction of rapid trains between some large European cities provides an example of the kind of innovation and international agreements needed.

A fundamental feature of metropolitan and other large urban regions relates to their role of being "novelty factories" for knowledge creation, the adoption of innovations, experimental interaction between customers and suppliers, the introduction of new lifestyles, the renewal of governance approaches, and the mobilization of resources for adjustment. They are arenas for face-to-face interactions and other proximity externalities, and fundamental innovations and technological change often have to be adopted in the urban context before they are diffused across a wider space.

A second basic feature relates to the observation that the contemporary natural environment is "man-made", or strongly influenced by actions from the human civilization. In an urban environment, this interplay between local ecosystems and constructed resource-handling systems is a predominating feature. Urban system management is challenged by the option to develop a tractable symbiosis of nature and culture in order to avoid "the tragedy of the commons." Two principal reasons make this difficult. Any major development project or institutional reform will involve a large number of urban stakeholders with different or conflicting interests, hence an agreement can be difficult to achieve. In the Stockholm region, representatives from at least some of its 26 municipalities, from the county council, and from the national government would be involved. Additionally, the politicians wanting to be reelected would have to seek public acceptance for a decision to support or not support an agreement. The political power game and referendum related to the introduction of congestion pricing provide ample evidence of these difficulties (Hårsman and Quigley, 2011).

If we look back at Stockholm's development, one point of departure would be to consider cities as the main drivers of income and wealth in Sweden and other countries. To take an opposite view, the global development of technology, trade, and commerce can be looked upon as a tide that has lifted Stockholm, as well as most other cities in the Western world. The two perspectives indicate the importance of considering the interdependence between city growth, national growth, and global growth when one is trying to understand the long-term development of a specific city such as Stockholm. The fact that Stockholm has increased its percentage share of the Swedish population dramatically and grown more rapidly than most other European cities since 1850 indicates that Stockholm has, at least to some extent and during some periods, functioned as a Swedish and also a European node of growth. But as will be evident from the following sections, this is not to deny that most of the growth impulses might have been external.

2.3 STOCKHOLM 1850–1914

An ugly duckling

Industrialization, urbanization, and efficient transport infrastructure came late to Sweden. As late as 1850, Sweden's largest city and capital, Stockholm, had just over 90,000 inhabitants and a constant birth deficit, so only a large influx of immigrants from rural areas prevented the city from shrinking. There were many small-scale craftsmen and some mechanized factories, but no large-scale factories. The city lacked water, sewerage, and gas lines. What waste management existed was hardly hygienic, and health care was substandard. Neither the city's inhabitants nor its own leaders had much influence over development issues; the Swedish Crown and the Swedish Church held the city in a firm grip. The rules governing commerce and trade were outdated and freedom of trade was limited, though the city monopoly on trading organizations and guilds had recently been revoked. For many goods, international trade was still hindered by bans on imports and exports, in stark contrast to free-trade-friendly Norway, despite the fact that Norway was also ruled by the king of Sweden (until 1905). Stockholm, until then isolated from the outside world during the winter months by harbor ice, could now extend its sailing season, thanks to the new steamships. But Sweden's capital city still lacked a railroad link.

When the Danish fairytale writer Hans Christian Andersen visited Stockholm in 1850, he wrote that this beautiful city of wide straits and high hills reminded him of Constantinople. This prompted one commentator to exclaim that the comparison was indeed valid: both cities had the same stench, dirt, and poor sanitation; both were ugly ducklings. But Stockholm was actually about to undergo a unique transformation into an international metropolis with a strong focus on the environment, sustainability, and quality of life. The city was, in other words, fairly backward in 1850. Environmental historians describing this period, however, note that even though the environment was unhealthy and damaging, most Stockholmers had a small ecological footprint: production, consumption, and waste cycled locally. That would change.

Political evolution

In 1850, Swedes looked with both envy and dread upon the modernization in Western and Central Europe – especially the United Kingdom. They saw in its local economies and cityscapes rising populations; young industries; steam engines, canals, and railways; and piped water, sewage, and gas. They saw cities that produced healthcare systems and a bountiful supply of food and consumer goods – but also poverty, housing shortages for many industrial workers and their families, growing gaps between rich and poor, political challenges, and social unrest. It was a new age – for better or worse.

Even in Sweden, agricultural modernization had increased productivity and profitability, and, together with better access to education through public school reform in 1842, this led to a larger and healthier population in rural areas, where 90 percent of Swedes still lived in 1850 and 80 percent by the century's end. Nevertheless, landless Swedes gradually migrated to the cities, where incomes and public health were also slowly improving, so both urban and rural areas across Sweden experienced population growth due to a surplus of births over deaths. Even during the period of the great exodus between 1850 and 1914, which saw one in five Swedes migrate to North America, the Swedish population grew by 63 percent. While rural areas grew by a third, the cities experienced fivefold growth – both contributing to a near-tripling of GDP (at constant prices).

The first stage of Stockholm's and Sweden's major transformation took place during the 1850s and 1860s. The new political forces of liberalism were in a constant feud with the still-strong conservative and protectionist interests, but nevertheless managed to enforce some fundamentally important national, regional, and local reforms. In 1862, (more or less) independently governed municipalities and county councils were given direct taxing powers, and by 1866 the parliament had two (more or less) democratically elected chambers replacing the antiquated parliamentary system of four estates composed of nobles, priests, burghers, and landed farmers. Other administrative reforms liberalized commercial and financial markets and instituted civil rights. The ground had thus been laid for the accelerating economic growth that would, during the century between 1870 and 1970, increase Swedish per capita GNP eightfold, an increase greater than in any other country except Japan. The overarching ideology of liberalism reigned: belief in the power of the market's invisible hand. But the liberal government stood ready to reach out its very visible hand when need be in the form of national support or regulation.

Simultaneous developments facilitated this rapid growth. A radical upgrading of the transportation infrastructure came with the introduction of railways. The raw-materials-based economy gradually became transformed into an economy dominated by technology-intensive production. And finally, cities could improve living standards by offering piped water, as well as sewage and waste systems.

Evolving infrastructure and urban densification

Sweden's first industries, based on rich resources of iron ore and timber, had grown in areas with access to water, power, raw materials, and shipping ports – none of which Stockholm offered, except a harbor. From the mid-nineteenth century, foreign demand increased for iron ore from central Sweden and timber from the north to supply the more industrialized nations' expanding manufacturing and building industries. The mercantilist export and import bans had been replaced first by variable tariffs, then, at the beginning of the 1860s, by comprehensive free trade – neither of which furthered Stockholm's relative position among Swedish ports. Industrialists used their profits to improve their export opportunities by building private railways between mines, factories, and harbors. The Swedish railway age was launched, and this primarily benefited two cities, Gävle in the north and Gothenburg on the North Sea.

But from the end of the 1850s, policies reflected a new acceptance that state capital imports should invest in the infrastructure of the future: railways, postal and telegraph systems. State-owned trunk rail lines financed by international loans linked state-backed, privately financed local railways, setting the stage for a nationwide railway network. Stockholm received its first rail links in the 1860s, and in 1871 these were linked together through the city to a central station near the central business district.

Stockholm's population doubled between 1850 and 1883, tripled by the end of the century, and quadrupled by 1913. Much of what is today Stockholm's inner city was built hastily, based on models from Berlin and Paris: five-story houses along long corridor streets and in backyards. Here and there, the compact environment was broken by broad, tree-lined boulevards and esplanades. Park plazas, with streets radiating from the center in a star formation reminiscent of the Place d'Etoile in Paris, were admired by artists and writers. The radical densification of the city, made necessary by the rapid population growth, was less spectacular. Small but nevertheless expensive apartments dominated the housing market; crowding and abominable sanitary conditions were a scourge. As time went on, however, an increasing share of apartments gained access to water, sewerage, and gas. The rural character of the outer districts of inner Stockholm, with their low wooden houses, kitchen gardens, and tobacco fields, was basically wiped out. But many hills remained, and these became "parks for all," starting a tradition of

Figure 2.1 Drawing from 1886 of Stockholm's central railway artery passing on a drawbridge across "the Sluice" (Slussen) and further across to the old city (left). Up to this day, this narrow passage, "the wasp's waist," remains the city's main railway connection southwards

Gustaf Broling, Stockholms Stadsmuseum Archive

public parks offering recreation, free theater, and other activities for all Stockholmers (Lundewall, 2006).

The Swedish Crown had long been the largest landowner within Stockholm's city boundaries and also the most adamant opponent of the City's ambitions to make new land available for housing. With crowding becoming severe in the inner city by the turn of the century, the conservative political parties within the city council launched a new strategy that would continue for the next 80 years: purchasing land outside the city limits for technical and social institutions as well as new housing and recreational areas.

Already, a few industrial suburbs offering substandard housing had cropped up near the railway stations, ports, and road crossings at the periphery, and in other areas private developers had begun to build new neighborhoods served by rail for wealthier residents, at a comfortable distance from the crowded city. A metropole was slowly taking shape. In 1913 and 1916, most remaining newly purchased areas of what we now call Stockholm were officially incorporated within city limits.[1]

The built environment in the inner city had become too large for most people to be able to walk (or row) to work. Rowboats were replaced by steam-powered ferries, and endlessly long stairs built into the hilly city were replaced by public elevators. Horse-drawn streetcars were introduced, and were a short time later electrified and integrated into a wider transit network. Cycling also became popular; in 1914 there were about 80,000 bicycles in Stockholm, but only a little over 3,000 automobiles and buses, and about the same number of horse-drawn vehicles.

Stockholm Harbor, which had long been the country's most important for both imports and exports, was expanded, but nevertheless lost market share for timber, iron ore, and steel to ports in the north and to Gothenburg. In 1850, 40 percent of Sweden's exports passed through Stockholm and 25 percent through Gothenburg. Thirty years later, the value of Sweden's exports had increased significantly, but Stockholm's export value share had sunk to 8 percent while Gothenburg's was unchanged (Hammarström, 1970).

As Sweden's capital, and the nation's wealthiest city, Stockholm provided the most important domestic market and was therefore able to defend the position of its harbor as the leading port for imports of consumer goods and production inputs to Sweden. This was especially the case for high-value goods, which would successively be strengthened even more, thus favoring the economic development of Stockholm.

Industrial evolution

The repeal, just before 1850, of the guilds' monopoly on manufacturing and trade, which had in part aimed at supporting the capital city's production and commerce, but also had protected obsolete structures, set free optimism and new ideas. And once the steam engines of the 1850s and 1860s became the city's most important energy source, Stockholm could develop into a modern industrial city. Industries established in the capital city grew quickly to become the largest in Sweden (Magnusson, 2010). Almost half of the wage-earning population of the city worked within industry or craft trades in 1860, and up until World War I this share decreased only slightly. The expansion of the city dramatically increased the number of industrial workers, as well as the volume and value of their output.

At first, industrial output was dominated by consumer goods for local markets: textiles, porcelain, groceries, beer and tobacco, newspapers and books. These were available in a large diversity, generated by small-scale producers, until they were outcompeted by larger companies engaged in the steam-driven mass production of both consumer goods and production inputs that could serve national and international markets. As the labor force became more skilled and labor costs increased, Stockholm's firms focused increasingly on finished consumer goods. Textile firms now offered clothing; metal providers became machine producers. Labor- and space-intensive industries were forced out of the inner city by the high cost of, and increased competition for, land, first to the suburbs and then farther afield. When electric motors and combustion engines were introduced, access to locally produced energy ceased to be a geographically limiting location factor (Hammarström, 1970).

Stockholm's industries also began to shift toward more high-technology innovation firms capitalizing on Swedish inventions and scientific advances. Many of these are still familiar names: L. M. Ericsson, Separator, Atlas, AGA, and Nobel.[2] Stockholm's industries also imported knowledge and skills, keeping a close eye on international developments, some supported by contacts with the Technology Institute (*Teknologiska institutet*), now the Royal Institute of Technology (KTH). A city of engineers and entrepreneurs could be glimpsed on the horizon.

The years around the turn of the last century would witness one of the most dynamic periods in the industrial history of Stockholm, and indeed the whole of Sweden. The steam age was replaced by the electric age. Sweden became an integral part of the international economy, and economic cycles swept in with greater force over Sweden's borders. Commercial banks, founded as rather humble entities in the 1850s and 1860s, had grown into leading actors in business development. The banks' influence, combined with the national government's economic initiatives and the practice of national agency procurement of the latest technology,[3] contributed to a transformation of domestic company clusters into internationally competitive corporations. Many of these still form the backbone of Sweden's manufacturing industry and have set their mark on Swedish society.

If this was a breakthrough age for the banks and industrialists, it was also the period when the labor unions began to develop, and with them the

Social Democratic Party, temperance leagues, independent churches, and other civil society movements. The estate-based society that had been abolished by the parliamentary reform in 1866 now became a class-based society composed of an upper class, middle class, and a new working class. In the decades to follow, conflicts among these classes were bitter and resolved only slowly. A new age was approaching.

Industrial Stockholm's combination of population growth, crowding, housing shortages, high rents, and higher average income during the "good times" before World War I led to an intensified and rejuvenated real estate construction boom, both on the scarce land in the inner city and in the newly incorporated areas. In central areas, existing neighborhoods were expanded, based on the same grid principle, but new areas reflected the influence of the English *garden cities*, with row houses transformed into free-standing, small, single-family homes in the Swedish tradition. Building permits were granted to less wealthy Stockholmers willing to build themselves (Kallstenius, 2010; Lundewall, 2006; Andersson *et al.*, 1997; Andersson, 2012; Eriksson, 1990; Johansson, 1987). Stockholm's political parties enjoyed a relative consensus regarding housing policy, but with different motives: from securing the political loyalty of the new dwellers to making owner-occupied homes accessible to families with lower incomes.

Improvements in sanitation and health

Waste management and recycling freed the city from putrid and infection-spreading open areas, so-called fly meetings, near market squares and other public places. The first steps were taken in 1859 with the inception of the Sanitation Administration (*Renhållningsverket*) to manage the collection of human waste from latrines; the waste was transported outside the city and converted to salable fertilizer. By 1900, the City of Stockholm had also taken on responsibility for cleaning the streets of horse manure and other pollutants, ameliorating the most serious sanitation hazards. Nevertheless, comprehensive waste collection would not be a municipal responsibility until 1972.

The first piped water systems from 1861 were built out and eventually all households had access to fresh water, which had previously been prohibitively expensive to all but a few. Stockholm had suffered several cholera outbreaks starting in 1834 but with successively less catastrophic effects; the last epidemic was in 1866. The sewerage standard was, however, considerably lower.

From 1853, the first gas production facility, gas lines, and new gas-lights meant that Stockholm was less directly dependent on daylight. Forty years later, gaslights would be electrified, but municipal gas production continued, as an increasing share of households replaced wood-fired cookers with electric ranges. Few households installed gas lamps, however; oil lamps

Figure 2.2 The waterworks at Eriksdal on Södermalm in central Stockholm, 1891

Stockholms Stadsmuseum Archive

dominated in Stockholm's homes for a few more decades until the introduction of electric bulbs.

Improvements to hygiene standards contributed to the doubling of
life expectancy for men from 25 to 50 years and for women from 37 to 51.
Improvements in the 1850s and 1860s, which continued through the balance
of that century, made life better for most Stockholmers.

The environmental cost of modernity

The *ecological footprint* of the city's inhabitants and industries, which had
been small in 1850, was somewhat larger by 1900. The rapidly expanding
population, land area, and industrial development had indeed contributed to
more consumption, more waste, and more transport trips. However, much
production and consumption was still cycled locally, and transportation was
still served primarily by horse-drawn carriages (Pettersson, 2008).

When gaslights and factory heating with coal or coke from the
gasworks were introduced, and when steam engines also began to burn
fossil fuels, air pollution became a problem, though not on the scale that the
automobile would bring. It would be many more years before the majority of
households discarded their wood-fired stoves or replaced fireplaces with oil-

fired central heating. Electricity was appearing in homes, industry, and transport systems, but was not yet widespread. Only later would electricity begin to replace fossil fuel imports – with the associated displacement of environmental effects – with domestically produced water power.

The first waterworks used a rather simple filtration system and were located in the vicinity of large sewage outlets into Stockholm's waterways, but many neglected the danger inherent in such a system, as Stockholm is surrounded by water and situated where one of Northern Europe's largest freshwater lakes flows out to sea. Experts argued in vain for sewage outlets to be located farther out from the city and for better water and sewage treatment systems. The cost of such systems discouraged policymakers, who instead preferred to encourage individual households to use charcoal filters to provide clean tap water.

Fortunately, this worked fairly well until Stockholm City Council decided in 1909 to make the installation of water closets more convenient by allowing them to be directly connected to sewage pipes, without the previously compulsory storage and sludge separation in septic tanks. The installation of indoor toilets became more common – and outflows of raw sewage became a new scourge for the city. Deteriorating water quality soon became noticeable and residents complained as public baths closed and fishing harvests dropped. Still, the first water purification plant would not open until the 1930s; it would be a long time before Stockholm's sewage would be purified mechanically, and even longer before purification would be extended over several phases.

A missed opportunity

Gustaf Richert, who was a professor in water systems architecture at KTH, was an early environmental activist who also served both on Stockholm City Council and in the national parliament around the turn of the last century. Richert not only passionately championed water purification but also voiced concern over increasing air pollution and proposed laws requiring concessions, control, and monitoring by a new public authority that could even take its own initiatives to reduce health risks. This eventually came to pass, but not until 50 years and many environmental accidents later. Both the national government and Stockholm's leadership had other concerns, *inter alia* with the outbreak of World War I, and Richert's proposals were essentially forgotten until Rachel Carson's *Silent Spring* awakened public opinion. Sweden's Environmental Protection Agency was created in 1967 (Strandh, 1985, pp. 11–13).

A recurring question in urban studies is the extent to which development is generated endogenously or, alternatively, exogenously. Stockholm's development until World War I was arguably most affected by driving forces at the national level, in turn affected by experiences and lessons learned from other countries. Finance Minister J. A. Gripenstedt (1813–1874), who had studied England's industrial districts and was inspired by the economics of "liberal harmony" personified by Frédéric Bastiat, was Sweden's most far-sighted politician in those days. He pushed through liberal reforms to the governance system and commercial law, brought about strategic investments in the railway network, and brokered cooperation between the national government and financial markets. He also facilitated national capital imports to finance the building of railways and the inception of commercial banks. Another leading figure was his friend the banker A. O. Wallenberg (1816–1886), who laid the groundwork for cooperation between banks and industry that has since characterized Swedish industrial structure (Ohlsson, 1994; Magnusson, 2010).

2.4 A PEACEFUL ERA OF SOWING SEEDS, 1914–1945

Many of the seeds of today's Stockholm and Sweden were sown in the latter half of the nineteenth century and around the turn of the century. However, the period called the "democratic breakthrough", in 1920–1921, was equally important. All adult citizens, both men and women, were given the right to vote and hold public office at the national and local levels. In Stockholm, the municipal governance system was reformed, simplifying earlier complicated and lengthy protocols. The city council was given decision authority, the Central Board (*stadskollegiet*) became the new governing body, and six city councilmen were assigned key areas of responsibility and associated committees (Larsson, 1977).

Neutral in two world wars

The twentieth century brought two world wars, and a number of civil wars, revolutions, occupations, and dictatorships to Europe. Only Sweden, Switzerland, and a few other small states were spared. Sweden's neighbors Denmark and Norway were also neutral during World War I, but were occupied by Nazi Germany during World War II. Throughout both wars, Sweden was cut off from its most important trading partners, Great Britain and the United States,[4] and its foreign trade was temporarily reoriented toward the German market. Sweden exported raw materials such as iron ore and other products that ostensibly preserved Sweden's neutral status and were not essential to the country's own defense or in the production of consumer products. Sweden's import and export volumes both shrank, as a large share of domestic industry reorganized so as to be able to replace imports with domestically produced goods; other firms slowed production as labor was rerouted to the country's defense services. Gender segregation decreased in the labor market, as many male jobs were assumed by women. Though they were not offered the same salaries or conditions, these were nonetheless important first steps toward gender equality.

The outbreak of war in 1914 broke the positive economic cycle that Sweden had enjoyed since 1910, and even up to the years just after the end of World War II the development of both agricultural and industrial productivity and profitability was weaker than it had been at the turn of the century. Neutral Sweden was not spared the hardships of the economic crises between the two world wars, which delayed the recovery of trade and goods production. By 1920, Sweden was almost paralyzed, though the crises of the 1930s were handled more effectively, and became less devastating.

Diversification and modernization of the industrial city

Stockholm continued its rapid population growth: the city population passed 400,000 during World War I and 500,000 during World War II. Simultaneously, the population of the surrounding county grew from 230,000 to over 300,000. Almost half of Sweden's population growth between the two world wars accrued in the Stockholm area.

As was noted earlier, Stockholm had begun to purchase and incorporate neighboring parishes in order to build housing for the growing population. And though World War I halted this progress and created a housing crisis for the poorest, planning continued so that building could recommence directly after the war (Johansson, 1987). Garden cities and multifamily housing were built on municipal land, as well as tramlines linking such housing areas to the inner city. As before, most multifamily housing was built by private developers but was increasingly often purchased by foundations and cooperatives and eventually by municipally owned housing companies, which raised technical and sanitary standards. Kitchens with gas or electric ranges and private bathrooms became more common, though by international standards the apartments still lacked many amenities.

Suburban commuter rail lines and buses allowed labor and housing markets to grow across municipal boundaries; a metropolitan region was now a reality for Stockholm's citizens, but not for its national and municipal leaders. The City of Stockholm was not part of the nationally governed regional county administration, nor did it come under the regional county council.[5] Many proposals to reform this administrative anomaly were proposed, but all of them failed until the 1960s.

Industry was still a powerful force in the reshaping of Swedish society. Sweden's transformation from an agrarian society to an urbanized country based on industrial and service sectors, begun in the nineteenth century, continued unabated. By the 1930s, the number of farmworkers was already lower than the number of those working in the industrial and building sectors, and a few years later it was lower than the number working in service sectors as well. Even so, about 60 percent of the Swedish population lived in rural areas, and it would not be until ten years after the end of the Second World War that the urban population constituted a majority.

The car and the airplane became popular, but mostly for commercial or government purposes rather than for private transport consumption. Horse-drawn carriages disappeared from Stockholm's streets, and electrified trams, commuter rail, buses, and bicycles were now the dominant modes of travel. Nonetheless, automobile ownership increased slowly but steadily. By 1939, there were 20,000 vehicles, one for every 23 people (Dufwa, 1985). The private automobile was more of a status symbol than an icon of personal freedom for working people.

Even though Stockholm was considered to be a major industrial city, services such as trade and communications had by World War I already begun to employ a larger share of workers. By the end of World War II, this share had doubled. Perhaps the most visible change in Stockholm's industrial character was that small and mid-sized industries, producing relatively unsophisticated goods, were replaced by larger, specialized, knowledge-intensive industries. This transformation would continue, and even be amplified, after the war.

The Social Democratic Party assumes control

Politically, democratic reforms marked the dawn of a long era of left-led dominance in Stockholm City Hall. From 1920 to 1950, the Social Democratic Party held the City's leading political post, that of vice-mayor of finance. Although the most prominent figures within the Social Democratic Party dominated Stockholm's leadership for many decades, the majority held by the Social Democratic Party varied considerably. Up until the 1980s, the largest parties shared the commissioner posts and the ruling of the city; thereafter, the opposition commissioners had no administrative respon-sibility.

The Social Democratic Party's dominance was at first not as strong at the national level as it was in Stockholm, although the party was Sweden's largest between 1917 and 2006, and led the national government for shorter periods before 1930 and then almost continuously thereafter: 1932–1976, 1982–1991, and 1994–2006, often with a majority in its own right but sometimes in multiparty coalitions. The first major challenge for the Social Democrats was the Great Depression, when mass unemployment gave rise to a spate of new initiatives, with the Swedish economic and political doctrine following a Keynesian approach even before Keynes published his *General Theory* (1936). New institutions and labor market laws were added, and employer and labor associations signed agreements that regulated negotia-tions and disputes. This was the breakthrough for the so-called Swedish model,[6] a set of principles guiding the division of power and responsibility

between labor market and state actors. Employers and unions would negotiate wages and other employment conditions independently, and the
government and parliament created laws and regulations, often in close
consultation with them. This construction proved successful and was applied
in other contexts: for the best result, let those most affected solve their
problems together.

Even though Depression-era Stockholm fared better than many
cities, unemployment increased dramatically, and the City invested in new
tram tunnels, bridges, and other types of relief work to provide jobs. The
central business district, which had developed in a neighborhood with a
street network from the seventeenth century and buildings from the
eighteenth and nineteenth centuries, was in urgent need of renovation.
An international competition gave inspiration to continued planning, but
the outbreak of World War II forced Stockholm to delay its investments.
Nevertheless, in 1941 the city council made a radical decision for a city of still
moderate size: to build a comprehensive subway network as the backbone
of a future metropolitan public transit system that would serve the city and
its suburbs.

This was a bold move, not only because it anticipated a development trajectory wherein the subway would play a central role in the
transportation system, but also because the City of Stockholm shouldered
the entire financial responsibility for such a large strategic investment
(Larsson 1977; Gullberg, 2001).

Splits in the environmental balance

The thirty-odd years between the outbreak of World War I and the end of
World War II was a time of enormous pressure for all of humanity, and though
less dramatic for those living in neutral Sweden and Stockholm, Stockholmers
did not go unaffected. The wealthiest regions in North America, Europe, and
Australia, and industrial nations such as Japan, dramatically increased their use
and misuse of ecological resources. Mobilization and armaments for the wars
squandered all types of resources: human, capital, and environmental.

The rapid population growth and improvements to the living
standard in Stockholm gave rise to debates regarding whether the capital city
was the engine of development for Sweden or a drain on its resources. This
debate would be raised often during the postwar period, but at this time was
focused more on spatial, socioeconomic, and population balance than on
environmental effects. Stockholm County had positive in-migration and low
birth rates, with many young people arriving from neighboring regions: from
1920 to 1940, Stockholm County's share of Sweden's population rose by
almost 25 percent while that of surrounding areas dropped (Schéele, 1991;
Snickars and Axelsson, 1984). This was not particularly problematic when
rural households were migrating to Stockholm seeking better opportunities,

Sök Er ej till Stockholm
21.000 SÖKER FÖRGÄVES BOSTAD

Figure 2.3 Poster produced by the City of Stockholm in 1946 to deter prospective immigrants from the countryside. The text reads: "Do not come to Stockholm: 21,000 are searching for accommodation in vain"

but when other cities began to lose population to Stockholm it was decried as unnatural and it was said that migrants had fallen victim to the seductions of big cities.

Wartime closures were barely noticed in the ever-rising curves describing the consumption of food, drink, consumer goods, transportation, and energy, not to mention the production of waste. Certainly, for a large share of the population, maintaining life's necessities was a challenge, but local public authorities mastered the art of rationing and developed effective systems that arguably even improved public health.

Except during the war years, Stockholm led the country in its increases in personal consumption, imported from other parts of Sweden and to some extent from abroad. Not only were food and drink imported, but the capital city was far from self-sufficient in most areas and thus exported more of its environmental effects than it had done previously.

The first sewage treatment plant was opened during the 1930s, and more would follow. The sanitation process was still rather crude, and rising consumption rates led to deteriorating water quality. Radical improvements would not be undertaken until the 1960s (Pettersson, 2008).

Electricity use had also risen dramatically across the country, and Stockholm doubled its consumption every decade. Gaslights had been electrified, and many households had replaced their gas ranges and started to buy electric appliances; the refrigerator replaced the icebox, electric lightbulbs shone in every room, and the radio was always on. Stockholm's first electric power facility was operational in 1892, but lacked a smokestack scrubber and was decommissioned after only nine years. Instead, a large electric power plant was built next to the gasworks in the harbor area, which could receive deliveries of feedstock fuels. This plant was built out and modernized successively, as demand increased and technology improved (Hallerdt, 1992).

Thanks to Sweden's hydroelectric resources and successively more technologically advanced and geographically extensive power grid, Stockholm began to abandon the use of imported coal. The City bought water rights and built power plants and transmission lines. Just after the end of World War I, City-owned hydroelectric power became available to Stockholm's consumers, and this pattern of acquisitions and expansions continued for several decades to meet steadily increasing demand.

As was noted earlier, Stockholmers did not begin to purchase private cars in large numbers until the late 1940s, so the automobile culture was not yet considered an environmental threat. The streetcar network had good coverage of the inner city, and many suburbs outside the city limit also had tram links with the urban core. There were also diesel-powered buses serving Stockholm, particularly in certain suburbs. In sum, the per capita environmental impact from transport was arguably moderate. The local and regional air quality also improved during this period, owing to the continued drift from local manufacturing to the service sector, which also led to a reduction in heavy transports for goods handling in the capital city area. From an economic perspective – and perhaps from a global environmental perspective as well – the continued concentration of power, management, and development of Swedish business in Stockholm was even more important. The shift of power from industry to the large banks that had begun at the end of the nineteenth century continued.

The period 1914–1945 was marked by change. Bitter class conflict gave way to peaceful conflict resolution between actors in the labor market and the evolutionary social democratic welfare state policy. The groundwork had been laid for a Swedish industrial society focused on technological development, large-scale production, and an export orientation. Stockholm developed in a similar vein, politically and industrially, and pursued urban development with modernism as the social and aesthetic beacon (Andersson, 2009; Eriksson, 2001). Stockholm's development was still heavily dependent on national driving forces and international role models, but there were already signs of the city's potential for endogenous growth.

2.5 REAPING HARVESTS AFTER 1945

Sweden renews itself, 1945–1970[7]

Joy at the news of Germany's unconditional surrender in May 1945 was as unconfined in Stockholm as in neighboring countries that had been occupied during the war. The general mood was buoyant and there was a sense of intense optimism for a peaceful future, as well as the urge to abandon rationing and consume again. This was stimulated by all the novelties now available on the market: American fashion and automobiles, Coca-Cola and hamburgers, plastic and penicillin. The government and many leading economists, however, feared that the Great Depression after World War I could be repeated if cautionary measures were not taken.

The victorious Allied forces had also prepared to speed up the wheels of economic development and avoid the famine and other horrors that had followed World War I. Payment and credit systems would be stabilized by Bretton Woods, the World Bank, and the IMF; global trade would be stimulated through tariff reductions and free trade agreements in GATT; and the Marshall Plan would rebuild Europe.

The result was rapid economic growth in the entire Western world – the Soviet bloc remained outside – and the Swedish government, instead of facing recession, was now forced to take strong anti-inflationary measures. Wartime restrictions, bans, and rationing were extended and expanded to dampen demand for inputs to industry and consumer goods to citizens. This produced a confusing and unpopular string of bilateral trade agreements and domestic regulations that would prove unsustainable and were eventually phased out, even though some parts would remain for decades; regulation of the construction industry continued until 1958 and regulation of the rental market has never been fully revoked.

From 1950 until 1965, the peak year for Swedish industry, when it employed 1 million workers, industry increased its productivity and profitability but also its energy consumption. The industrial growth rate was still high in the late 1960s, but from 1970 on it was only half as large. This golden age reflected not only global trends but also Sweden's definitive transformation from a relatively underproductive agricultural and forestry economy to a highly productive economy based on manufacturing and services.

Swedes became healthier and lived longer. In 1945, the life expectancy for men was 66 and for women 71; in 1970, men could expect to live to 70 years and women to 77. Swedes were also materially wealthier and income disparities shrank. Total consumption doubled between 1950 and 1970, but over half of this growth was due to tax-financed public consumption: health care, nursing, education, public administration, defense, and prisons.

From 1950 to 1970, GDP doubled (at constant prices), but municipal expenditures tripled, from 12 percent of GDP in 1950 to 18 in 1970. Municipalities were strengthened in two ways. First, two amalgamations reduced the number of municipalities from 2,300 to 290; and second, the scope of municipal responsibility increased dramatically. Stockholm County now comprises 26 municipalities, but the City of Stockholm is by far the largest, with 40 percent of the county's population.

During these two decades, many small farms in Sweden's sparsely populated areas closed and farmers and workers migrated to cities and towns across the country to work in industry and the public sector. The cities and larger towns were building modern, sanitary apartments as part of the "Million (Homes) Program," and there were better opportunities for working women in particular. Some of this migration went to the largest cities, but not to the extent that many feared it would.

Sweden, formerly a country of net emigration, was now a destination for immigrants as industry attracted labor from Finland, Poland, and Southern Europe. This was new for Sweden; the country had welcomed refugees during wartime, most of whom were already assimilated, and new waves of immigrants arrived in the 1940s and 1950s. Industrial towns became more ethnically diverse. Swedes in general became more tolerant of other cultures and lifestyles, and started to travel abroad more often. By the end of this period, however, labor immigration was reduced by national policies. Instead, refugees and their families from outside Europe began to migrate to Sweden, making the word "multicultural" (*mångkulturell*) a living part of the Swedish language and culture: a badge of honor for many, a negatively charged term for others.

Renewing Stockholm, 1945–1970s

The transformation of Stockholm from being an important base for Swedish industry, to being almost exclusively dominated by private and public services, continued unabated. The inner-city environment was particularly affected as manufacturing moved out, changed in character, or disappeared.

The few manufacturing industries that remained were in graphic services, specialized groceries, handicrafts, and specialized manufacturing of proto-types for demonstrations. Within industrial corporations, management and development functions became increasingly separated from production and logistics, and preferred localization in the Stockholm region, particularly the inner city. As a consequence, the area became even more attractive for all sorts of business service and consultancy agencies. Stockholm was on the verge of becoming a global metropolis (Hall, 1998; Magnusson 2010).

Urban development planning was enjoying a renaissance and found new forms: master plans and regional plans. Large cities were the front line

Figure 2.4 The square in Vällingby, internationally renowned "model suburb" built according to the ideal of the so-called ABC (Arbete, Bostad, Centrum = Workplace, Dwelling, Centre) city

Lennart af Petersens, Stockholms Stadsmuseum Archive

of development; Stockholm incorporated new outer-city areas in both the northwest and the southwest. Stockholm's Master Plan of 1952 outlined the guidelines for transit-oriented urban development, with newly built model districts laid out like a string of pearls along subway corridors (Stockholms stads stadsplanekontor, 1952). One of these districts, Vällingby, became internationally famous (Sax, 1998).

The 1952 Master Plan was actually never formally adopted but nevertheless laid the foundation for planning and urban development for decades to come. Stockholm built as never before – or since! In the urban core, the central business district underwent a radical redevelopment around

Figure 2.5 The redevelopment of the central business district began in the 1950s with the construction of a road tunnel and three intersecting main subway lines, which had to be built from the surface and required extensive demolition of a number of central quarters, where the present Sergels Torg (Sergel Square) is situated.

Lennart af Petersens, Stockholms Stadsmuseum Archive

Stockholm Central Station, where several subway lines linked to the commuter, regional, and national trains. Stockholm's modern amenities were to be easily accessed from the modern new districts at the periphery (see Stockholms stads generalplaneberedning, 1963, 1965).

The construction of the subway lines began in 1944 and continued for the next 50 years. This was also the start of a large-scale urban redevelopment that would last 20 years (Sidenbladh, 1985; Gullberg, 2001). The greatest changes – demolitions, provisional solutions, new constructions – tested worried residents' patience. A growing number demanded that these radical projects be halted and called for a return to a more cautious approach to refurbishing the city. City leaders gave in to public opinion, and planning in the 1970s reflected a new paradigm favoring preservation and traffic calming. This also implied considerably less strain on municipal coffers.

Simultaneously, Stockholm's population growth began to slow and change in character. Higher household incomes resulted in new lifestyle choices; a growing number chose larger domiciles closer to green areas, commuting to the inner city from the suburbs. As traffic increased, the city population shrank; by 1980, Stockholm had 647,000 residents, 160,000 fewer than its peak population in 1960. In the meantime, the metropolitan population outside of Stockholm more than doubled, from 358,000 to 740,000.

In 1963, a century after county councils had been established in Sweden, Stockholm's vice-mayor for finance proposed that the city apply to join the Stockholm County Council and coordinate the planning for the entire metropolitan region, including land use, public transit, and public health care. In 1971, the new, enlarged county council became operational. Three years earlier, the County Administrative Board had replaced the former Local Administrative Board of Stockholm (Wijkmark, 2002). One responsibility of Stockholm County Council was the intraregional redistribution of tax income and the provision of expansion loans to fund the building of new housing estates in less wealthy municipalities, although this was eventually phased out.

The provision of housing continued to be a primary municipal responsibility in the region and had been ratcheted up to avoid renewed housing shortages. As urban population declined, the region was suddenly faced with a glut of public housing apartments that would fill with tenants only slowly – some with refugee immigrants from outside Europe – which led to ethnic segregation, weakening municipal revenues, and increasing critique of the Million (Homes) Program.[8]

Household relocation and commuting moved farther and farther afield. The coordination of regional planning was a natural response to this phenomenon, as population growth was now taking place less in the urban core and more in the surrounding region. An Outline Regional Plan from 1966 anticipated that Stockholmers' recreation area, and before long also the labor

Figure 2.6 Aerial photograph of Rinkeby at Järvafältet, one of the major developments of the Million Homes Program
Ingrid Johansson, Stockholms Stadsmuseums Archive

and housing region, would spill over the borders of Stockholm County (Wijkmark, 2002).

This outline plan was based on expectations of strong regional population growth and a significant increase in living standards; therefore, it included more housing areas and transportation infrastructure – both railways and roads – than previous plans had. The plan provoked strong criticism, not least from environmental interests, and was never formally approved. Most of its ambitions would not be realized at that time, particularly new road projects, for example an eastern orbital link (currently again discussed) and a western bypass of the urban core (currently under construction). Many of the housing areas proposed by the plan were reconceived as nature reserves, or were downsized or postponed. On the other hand, proposals in the plan for significant improvements such as the development of large-scale regional sewage and waste treatment facilities were in fact implemented during the 1960s and in the decades that followed, and these have played a major role in establishing Stockholm's reputation as an environmental role model. Nowadays, most waste is recycled or combusted to produce energy,[9] and construction waste has been used to make man-made hills for winter sports. Stockholm's water is now so clean that it is suitable for fishing and bathing almost everywhere (Ingo, 2002).

Dreams of an endless supply of energy

The correlation between living standard, comfort, and energy use is well established. Stockholm experienced this at the end of the nineteenth century with the introduction of electric lighting. Thereafter, electricity use in Sweden doubled every 11 years, with an even more rapid growth in Stockholm, and production facilities were built out rapidly. In the mid-1960s, hydroelectric power provided most of the electricity used in the country. The rest was imported; cheap and easily handled imported petroleum had almost out-competed coal and coke. Still, Sweden's total fossil fuel use was significant, and air pollution from transportation and smokestacks was on the rise.

Increasing the use of electric power seemed in this perspective to be a solution to the negative consequences of imported fossil fuel, and electric power producers forecast dramatic increases in demand. A large share of Swedish industry, including the steel and paper pulp industries, was (and remains) a large net consumer of electricity. Electricity demand also increased in order to power railroads and streetcars, not to mention the steadily increasing household demand. New uses for electricity were being introduced every decade, and they came to the cities first.

Electric power plants in Stockholm had also become more efficient and more profitable by providing not only electric power but also district heating to all new housing areas on municipally owned land – indeed, almost all new development within city limits. This provided building owners with a more environmentally friendly, cleaner alternative to individual oil burners – not to mention freeing up the space they took. The big news was nuclear power, which many saw as an unlimited resource. In his speech to the United Nations in 1953, "Atoms for Peace," US president Dwight D. Eisenhower had proposed the establishment of the International Atomic Energy Agency and offered all countries willing to submit to its control access to material and knowledge about the peaceful utilization of nuclear power. Sweden and Swedish industry were early adopters, not least in Stockholm, where test reactors were built on the KTH campus and in Vällingby's counterpart on the southern edge of the city, Farsta, which would be fully supported by nuclear electricity and heat for 11 years. Both reactors have now been decommissioned and nuclear power is a controversial energy source to say the least, owing to its serious risks and the difficulties posed by the managing of nuclear waste (Hallerdt, 1992).

A symbol of progress and freedom becomes an environmental villain

If we look back, the doubling of Swedish consumption in the space of two decades was not only positive. New durable consumer goods – automobiles, refrigerators, freezers, recreational boats, televisions, and much more – made life richer and more comfortable, but also had both public and private economic consequences. Above all, the automobile demanded large investments in streets and roads; petrol import, refining, and distribution; service stations and junkyards; as well as additional police, ambulances, and accident response teams. For the Swedish automobile industry and importers of cars and petroleum, this was a golden age.

Most, especially men, still considered the car a symbol of freedom. Even if public opinion decried the congestion and accidents that increased automobile use entailed, they were not well aware of environmental threats from emissions and particulates. And few had any idea that pollutants affected areas beyond an immediate vicinity. Combined with the increasing use of oil heating in houses and facilities, automobile use began to threaten the earth's climate. Today, this is common knowledge, but then it was understood only by experts – and they were often met with skepticism.

Despite Sweden's large surface area, long distances between cities, and large transportation demands, neither the road nor the rail network was particularly well developed or of a world-class standard, apart from the major road and rail corridors: trunk routes, highways, and European motorways. Since the major extensions between 1857 and World War I, there had been few large investments in new railways, maintenance was substandard, and service on many lines was discontinued. This negative spiral accelerated after World War II, as the focus turned to the developing of highways. Stockholm had a better local and regional public transit system in terms of capacity and access than other Swedish regions, accounting for almost half of trips to the urban core and half of the entire country's transit trips. Meanwhile, the road network was extended, but could not keep pace with the expanding car culture.

Still, notwithstanding the speed and flexibility offered by the car compared to the more restricted transit network, a wave of public opinion in the 1970s voiced criticism over the rapidly developing automobile fleet and new roads, more so than in many large Western cities. A partial explanation may be that local transportation policy in Stockholm had in the 1950s and 1960s focused on expanding both railways and motorways, but did not simultaneously invest in infrastructure for walking and bicycle trips, as Denmark did. There is essentially no mention of bicycle infrastructure in Stockholm's planning and background strategy reports before 1973, nor are cycles given much mention in central planning documents. Stockholm still lacks a cohesive and safe cycle lane network, even in the inner city. This may be why current debate tends to pit car use against cycling. Differences in transport mode share by gender and class have not made it easier to formulate and implement a more balanced transportation policy.

When Stockholm City Council decided in 1941 to base public transport on a subway system, Stockholm was in no way an automobile-dominated city. Most of the 20,000 cars registered in Stockholm before World War II had been decommissioned because of the war. Most believed that Sweden and Stockholm after the war would follow America's example

Figure 2.7 The public art in Stockholm's subway system is world renowned, such as Ulrik Samuelson's sculptural installations in the Kungsträdgården station

Hans Ekestang, Stockholm Public Transport/ SL:s bildbank.

and become more car-intensive, but few realized the extent to which private automobiles would soon dominate Swedish cities. In hindsight, though the subway in Stockholm was built to serve social and economic rather than environment and climate goals, we can be grateful for the city's foresight and the willingness of that generation to shoulder the entire investment costs in the subway system.

In the first decades after World War II, the spatial pattern of the region's built environment began showing signs of becoming transformed from a monocentric to a polycentric metropolitan form, even if the City of Stockholm was still the region's undisputed core. The industrial city gave way to the city of private and public services with an increasingly evident focus on leadership functions, research and development (R&D), and innovation. Stockholm had its own networks of cooperation and competition with metropolitan regions abroad. For the first time, Stockholm could be described as a network society with at least some endogenously generated growth. City politics was no longer dominated by a single party but rather by several strong political actors. Higher living standards and educational levels were reflected in more sophisticated citizen and consumer demands and an emerging culture of individualism.

2.6 THE PAST 40 YEARS

A growth crisis from 1970 . . .

In the years around 1970, new concerns arose as youth across the Western world protested against the blind faith of the previous generation in its infallibility and the refusal of that generation to take responsibility for serious global problems: environmental destruction, poverty, famine, and oppression. As a consequence of these concerns, Sweden took new initiatives in several areas, including the "1 percent goal" for foreign aid,[10] and a more vigorous critical opposition to oppression and colonialist wars. The Swedish government also encouraged the United Nations to announce an international environmental protection conference, which was held in Stockholm in 1972 and resulted in the Stockholm Declaration on the need to protect the global

Figure 2.8 The 1971 "battle of the Elms" protest against a planned subway entrance in Kungsträdgården in Stockholm's central business district, marking the end of the comprehensive urban redevelopment of Stockholm and the awakening of a broad environmental consciousness in the city

Björn Gustafsson, Alternativ Stad

environment, as well as the creation of the United Nations Environment Programme (UNEP). These policy successes gave Stockholm a (perhaps not wholly deserved) reputation as a role model in environmental policy: the conference was as much of a "wake-up call" in Stockholm as in many other cities. Stockholm's environmental reputation may be somewhat better motivated today, when Stockholm has an active and broad environmental movement, and when most cities, not least Stockholm, have given environment and climate issues a central role in local policy.

The capital city and region's rapid economic and population growth in the 1950s and 1960s also provoked a reaction from national politicians, but this came too late and was arguably misguided. A number of regional investment policies aimed at dampening growth in Stockholm, which was seen as draining potential from struggling regions in the rest of Sweden. This policy initially had positive effects for prioritized regions; in the 1970s, new universities and colleges were established as supportive anchors for regional development, and major national administrations, agencies, and military complexes were relocated from Stockholm to other parts of the country. An unanticipated side effect of this policy was that the private sector's share of employment in the Stockholm region, already high relative to the rest of Sweden, increased even more and strengthened its potential for economic expansion.

Sweden's ambitious international engagement could not mask growing domestic problems. During the last three decades of the twentieth century, a number of key export industries collapsed, including shipyards, iron ore, and steel. And domestic industries began to fail in the 1960s: textiles and clothing, leather and shoes, and rubber. This had negative consequences for many parts of Sweden, but not for Stockholm. Instead, Stockholm suffered from the effects of population decline; the construction industry and its subcontractors lagged and municipal budget revenues dipped.

At the same time, positive changes were evident in the labor market, especially in cities and, most of all, in the largest metropolitan regions. Family and gender equality policies made advances toward the emancipation of women through individual tax declaration reforms and the development of a comprehensive subsidized childcare system. These reforms gave working women previously unimaginable opportunities to find jobs in the labor market and earn their own incomes. The share of women aged 16–65 who were gainfully employed rose from 58 percent in 1970 to 80 percent in 1990.

But the famous successes of the Social Democratic Party's national economic policy, previously able to simultaneously achieve a high growth rate, full employment, stable prices, and a satisfactory balance of payments, no longer worked as it had before. For the first time in 40 years, Sweden elected a national government led by a party other than the Social Democrats, and many county councils and municipalities followed suit. A new constitution in 1974 replaced the two-chamber governance system with a single-chamber parliament, and common election days for parliament, county, and municipal councils were instituted. Ever since, Sweden's politics have been more like those in other parts of the West, with shifting political majorities in power. Both right-wing and left-wing winds sweep the country, and public opinion has a more immediate impact on policy making.

The national government elected in 1976, and following cabinets as well, were forced to devaluate the currency several times to retain Sweden's competitive position and high standards, but the domestic economy was too small to face the effects of international crisis: fourfold oil price shocks, economic warfare between East and West, nuclear power plant accidents in the United States and Soviet Union, and distant wars with global economic repercussions.

But all was not bleak. Despite the oil crisis, Sweden had inexpensive energy for its households and industries thanks to its nuclear and hydro-electric capacity. In sector after sector, the march to oil independence was on – and the country marched toward a new dependence: on nuclear power! Hydroelectric capacity could not be increased much more without harming fragile ecological areas and natural values such as untouched rivers and mountain lakes, and there seemed to be no other large-scale alternative that did not harm air quality or other environmental areas.

The oil crisis, combined with concern about oil dependency, has-tened a transition to direct electric heating of single-family homes. Electricity was clean, comfortable and – at least locally – pollution-free. In the Stockholm region, this phenomenon was particularly widespread during the 1970s. Stockholm used more hydroelectric energy than ever, but use of fossil fuels was still considerable as decision makers awaited the construction of a nuclear power facility in the region or access to hot water from existing reactors. But neither of these situations came to pass, because a public referendum in 1980 led to a policy aiming at eventually phasing out nuclear power. The country still remained oil-dependent, even if this dependence was now disguised as clean electricity.

. . . and rejuvenation

Despite foreign crises that affected Sweden, and despite the public referen-dum against nuclear power and flare-ups of ideological conflicts regarding socialization of industry, Sweden maintained the critical capacity for peaceful cooperation across party blocs and with the labor market actors that had developed in the 1930s. Consensus reforms in the 1990s focused on the taxation system and the budget process. Sweden's public finances were thus more stable by the turn of the century than they had been in 1970. They still are.

Another legacy of this strength was, ironically, the dual structure of Sweden's industry and exports: partly, this structure was based on and located near raw materials from most regions in the country (except the

Stockholm region), and partly it was based on service- and knowledge-intensive products from Stockholm and other major metropolitan regions. Sweden proved that, despite its smallness, it could keep pace with – and even lead – the rapid technological advances in areas such as information technology and life sciences, and utilize a long tradition in Swedish industry that focused on innovation and technological development.

In Sweden's larger cities, these transformations were manifested on such a large scale that they could be described as Schumpeterian creative destruction that forced out traditional manufacturing but ushered in a new technological age (cf. Schumpeter, 1943). Åke E. Andersson (1985) describes this period as Sweden's *fourth logistical revolution*, wherein the driving forces of knowledge, creativity, and communication first transformed the largest metropolitan regions and eventually spread throughout the country. In Sweden's second and third largest cities, Gothenburg and Malmö, large, centrally located shipyards were razed to make room for new types of industry, services, and housing. And in Greater Stockholm, the newest urban district, Kista, had a profile previously unknown to Sweden: an ICT and electronics cluster. Kista surprised both urban planners and politicians, who had anticipated a typical subway transit-serviced urban district but instead got two companies within the Ericsson group as the first leaseholders in 1976, followed by IBM – and then a host of other firms with similar profiles, one after the other. The snowball had begun to roll and Sweden's first high-technology cluster appeared.

Stockholm's reactions to these phenomena were reflected in two industrial and labor policy programs in 1976 and 1982. The first was focused on stimulants and support to counteract the disappearance of manufacturing firms from the city. Universities and research institutes were hardly looked upon as partners in the 1976 program, but in the second program they had assumed a role as important actors in local and regional innovation systems. The 1982 plan names electronics and other types of micro-technology as particularly promising.

At around the same time, and in coordination with the City's policy, Stockholm County Council assumed a greater responsibility for strategic development in the region. In particular, it actively supported public health and clinical medical research at the university hospital in the southern part of the county. This strategy served several goals simultaneously: it supported strategically important industrial sectors, developed the region's leading medical teaching hospital and medical school, and created life science research institutes, as well as a new teaching and research institute. All these measures led to the clustering of research in the humanities, social sciences, and natural sciences in the same area (Wijkmark, 2002).

At first, many of the City and County initiatives seemed to be conflicting, but eventually the degree of cooperation deepened and broadened, taking in several of the region's institutes for higher education and other municipalities. In its 1982 program, the City of Stockholm noted that a rich and diverse industrial base throughout the region was in the city's interest, and also that cooperation with other municipalities was improving.

The background to both programs was concern over the future of the city and the region. In 1976, manufacturing sectors were faltering and many private companies had serious profitability problems. On the other hand, the high proportion of women in the labor force was a positive sign.

By 1982, population and economic indicators once again began to point upward. Growth in Sweden's largest and densest region, with its rich supply of skilled labor, advanced consumption profile, and good access to knowledge-oriented jobs, has been strong ever since. The region is considered powerful and competitive. Several indices of Stockholm's competitive position in Europe and within the OECD show a region holding its own, and underscore success factors including high rates of entrepreneurship, a high rate of innovation and renewal, and significant public and private investments in research and development. No other Swedish region has such a diverse industrial base. As might be expected, Sweden's second and third largest regions follow, but the industrial base in the rest of the country is far less diverse. Stockholm is home to a significantly higher share of total labor force in the knowledge industries; the share of the workforce in ICT is twice that of other Swedish labor market regions (Johansson and Strömquist, 2002).

Stockholm's labor market development prospects are better across the board than in the rest of Sweden, and the region's share of GDP has risen faster than its share of population. All in all, the region has fared generally better than other Swedish regions in the face of economic downturns and has a proven economic resilience.

A cosmopolitan capital

The ICT cluster in Kista is often described as a symbol for the renewal of Stockholm's economy, but perhaps it rather reflects the accelerating globalization of information, economies, and societies – and Stockholm's entry into the circle of internationally recognized metropolitan regions (Linzie, 2002).

In part because of Sweden's liberal immigration policies for political refugees, the population of the Stockholm region became more diverse and daily life more cosmopolitan. Stockholmers traveled abroad and received foreign visitors more often than before. Growing tourism made its mark on the urban environment: giant cruising ferries began docking at Stockholm's harbors, and domestic hotel chains were sold and international hotel chains appeared in their place. Pizzerias, pubs, and hamburger joints sprang up and stores switched the Swedish language signs announcing a "*REA*" for the internationally recognized "*SALE*." Entertainment life glittered, business life was more globalized, and English became the language spoken in many companies and universities.

Stockholm's traditional role as the central node in Sweden's road and railway network extended to cover airlines, telecommunications, and data traffic. Arlanda became a major national and international airport serving over two-thirds of Sweden's air travelers, of whom 80 percent had an origin or destination in the Stockholm region.

Politics also underwent renewal: both the City and the County were determined to develop on the basis of a wider array of perspectives, to act within European and international arenas, formulate visions and goals for the future, and promote Stockholm and compete globally.

A new national Planning and Building Act from 1987 mandates long-term, comprehensive municipal physical plans. For the first time since 1952, Stockholm renewed its master plan and has updated it regularly ever since. These later plans have a common focus on densification of the inner city and the near periphery in order to better utilize centrally located land, improve the preconditions for public transit and cycling, and counteract sprawl and private car dependence (Ingo, 2002). In each new planning process, it is ever more evident that such ambitions are part and parcel of an active environment and climate policy approach to physical planning – even if plans offer, in practice, less room than earlier for the traditional idea that today's Stockholmers should leave to coming generations a "green city" in the strictest sense of the word. This new planning paradigm has also given form to new showcase neighborhoods such as Hammarby Sjöstad and the Royal Seaport – in spirit similar to Vällingby in the 1952 Master Plan, but better suited to current sustainability goals.

The aspect of city planning policy that aroused the most difficult conflicts and impasses during this period – in Stockholm as well as Sweden's other two metropolitan regions – was the development of the transportation system. In 1990, the national government commissioned three experienced and respected negotiators to facilitate discussions within each metropolitan region regarding a sustainable transport investment program that could harmonize needs for environmental protection with industry's transportation needs. In Stockholm, former Bank of Sweden director Bengt Dennis was given the responsibility to oversee this process, and after many rounds of negotiations the three largest political parties agreed on what came to be called the "Dennis Package" of transport investments covering the period 1991–2005. Despite considerable negative local public opinion, most of these transport projects have since been implemented.

Industrial policy was far less controversial. There was a general consensus that Stockholm should strive to be a diverse, international metropolitan region – a society characterized by Knowledge, Competence, Creativity, and Communications (see Andersson, 1985). In 2007, the Stockholm City Council adopted a 2030 vision for a growing, innovative region of world class and Stockholm began marketing itself as "the Capital of Scandinavia," and the county adopted a vision of itself as "Europe's most attractive metropolitan region." Realistic or not, these slogans reflect ambitions that require well-constructed policies for both the city and the region to ensure a good living environment for residents, businesses, and visitors.

Since the renaissance of economic and population growth, most of what has happened has been positive for the capital city region. Stockholm's population continues to grow annually by some 16,000 in the city plus another 20,000 in the outer parts of the region, and the pace of growth is, so far, stable. Stockholm has a surplus of births over deaths, as well as a surplus of both domestic and foreign in-migration over out-migration. Still, for many years foreign net migration has represented almost half of the total net migration for both the city and the region. As it is rather sensitive to changing circumstances and politics both in Sweden and abroad, this fact could serve as a memento for those who make projections and plans.

Perhaps even more important for Stockholm's future is the continued geographical extension of the functional labor and housing market regions to include large parts of adjacent administrative regions. Transport infrastructure expansions have reduced commuting times, and more people also commute over longer distances. By international standards, the capital region is still relatively sparsely populated in large parts of its area, but for Sweden's late-to-urbanize population open areas and access to nature are valued highly. Simultaneously, cooperation among the County Administrative Boards, county councils and municipalities has expanded to cover a range of issues affecting most of east-central Sweden. The capital city region's many networks are becoming knitted together, and municipal and county borders are becoming less relevant with every passing year.

The sustainable capital

No other policy area has been so expansive in Stockholm over the past decades as environmental policy. This reflects not only better knowledge regarding environmental challenges and climate change threats, but also the fact that environmental issues affect us all: individuals, households and companies, municipalities, regions, countries, and supranational organizations. Political parties, public authorities, scientists, civil society organizations, lobbyists, and the media are all touched by these issues.

Awareness of threats to environmental sustainability has grown successively in Sweden, as in other countries. After Rachel Carson's *Silent Spring* (1962) set alarm bells ringing with warnings of the spread of toxins,[11] environmental issues were raised in area after area: new crisis reports, new environmental catastrophes, new research, new policy, new laws, etc. Goals and budgets for environment and public health authorities increased many times over, and focus shifted from specific sectors to understanding the functional links and effects of complex systems.

The 1972 Stockholm Conference had shone a light on Stockholm and inspired municipal and regional authorities to increase their activities related to environmental protection. There – as in most parts of the world – the phrase "sustainable development" from the Brundtland Report, *Our Common Future* (World Commission on Environment and Development, 1987) became an expression of the common good, and the 1992 *Agenda 21* from the Rio Conference was even more influential at the local and regional scales. New concepts were introduced and new branches of environmental policy developed, of which climate policy, underscoring the interdependence of all peoples and nations, would seem to be of paramount importance.

The City and County of Stockholm have environmental programs focusing on goals for their own activities and the ways in which public awareness and conduct can be affected by information and programs produced by public authorities. In the transport and energy sectors, much has been done, but much remains if this active and successful metropolitan region, with such a high material living standard, is to meet ambitious national sustainability goals.

Stockholm has recently decided to link environmental programs to the city's economic planning. This decision has resulted in a common policy for socially, economically, and ecologically sustainable urban development. Stockholm has also entered into common agreements with other capital cities through platforms such as the *Covenant of Mayors 2007*, wherein signatory cities pledge to work for a more rapid reduction of greenhouse gases than that mandated by the European Union (currently 20 percent in the period 1990–2020). According to Stockholm's current action plan for climate and energy, already implemented and planned initiatives should be sufficient for the city to reach its emission targets for 2015.

The City's direct involvement in one of the most important areas for environment and climate policy, namely energy provision, was, through privatization, phased out partially in 1991 and completely by 2001. The City thereby lost direct control over the production and distribution of gas, electricity, and district heating, and therefore could no longer steer choices between renewable and fossil feedstocks, or use pricing policies to affect consumption and its societal effects. The environmental importance and complexity of the energy system acts as a reminder of another complex system for which privatization is under debate: public transit, for now still publicly controlled by the Stockholm County Council.

Growing car traffic – and concerns about the environment and the climate – prompted the Swedish government in 2002 to finance a six-month full-scale trial of a congestion charging system, set up as a special national tax with a charge cordon around Stockholm's inner city in combination with temporary expansions in transit services. After the trial, residents were invited to vote in a local referendum on a proposal to make the trial permanent. Local political parties were split regarding this new element of Swedish taxation policy, which had already been successfully implemented in the shape of traditional tolls or time-differentiated congestion charges in Singapore, London, and several of Norway's larger cities. The road charging trial in Stockholm was launched in 2006 and made permanent in 2007 after having passed the local referendum. Stockholm's political parties and the national government have since agreed to keep the congestion charges (or rather congestion taxes), even if they are divided regarding the use of revenues: to improve roads (benefiting those who pay the taxes) or public transit (providing an alternative to automobile use) (Isaksson, 2008).

As was discussed in the previous chapter, the European Commission named Stockholm Europe's first Green Capital in 2010. The award can be seen as a recognition not only of the city itself but also of city and county residents, organizations, and companies. Everyone's participation has contributed to the goals achieved, and continued engagement is needed to meet future goals. Inspired by the positive international attention, Stockholm's political leaders have established a new goal, namely that the city will remain Europe's environmental capital even after 2010. That is a challenge.

2.7 CONCLUSIONS: LOOKING BACK AND LOOKING FORWARD

Between 1850 and 2010, the population of Stockholm grew from 93,000 to 847,000, and close to 1.6 million people were added to the population of the surrounding, successively expanding commuting region. The corresponding share of the Swedish national population living in the Stockholm labor market region increased from around 3 percent to slightly more than 25 percent. This development represents a higher growth rate than the average among European urban regions. The population size rank of Stockholm's contiguous built-up area is currently 30 as compared to around 45 in 1850. Investments in the transport infrastructure help explain the rapid and sustained growth. The railway investments in Sweden during the last half of the nineteenth century increased the accessibility of Stockholm, and later investments conserved and strengthened Stockholm's role as a central node in the Swedish transport and communication network. Later on, economies of scale, growth of the service industries, and an increasing knowledge orientation would seem to be more important explanatory factors.

Looking back 160 years, Stockholm's economic growth might seem more impressive than recent achievements in reducing greenhouse gas emissions and improving the local environment. However, an observer from 1950 might have come to the opposite conclusion, since improving the local environment had tangible and direct benefits to residents. By 1950, successive investments in the water and sewage systems, in garbage handling, and in better housing had contributed to a dramatic increase in life expectancy: from 25 to 67 years for men, and from 31 to 72 for women.

A political scientist would perhaps be more impressed by the establishment of a modern system of governance and by the social engineering and pragmatism characterizing much local and national policy since the Social Democratic Party began to dominate the political arena some 80 years back. The reformation of Stockholm's public administration in the 1920s, municipal reforms in 1952 and 1970, the transfer of responsibilities from the municipal tier to the county initiated in 1963 and operational in 1971, as well as the congestion charging system in 2007 are striking examples of this pragmatism. The political scientist might even argue that the long-used practice of applying a problem-solving attitude and looking for solutions that can be accepted across party lines can now be considered important institutional capital – a capital that helps feed the political process with new knowledge and reduces the risk of political deadlocks – and that this might help make Stockholm still greener.

However, Stockholm's recently adopted action plan for reducing the emission of greenhouse gases indicates that this institutional capital might have eroded somewhat. Though ambitious in most respects, the plan adopts a rather narrow perspective by only considering the emission of greenhouse gases produced within the city border. According to Axelsson (2012), this definition grossly underestimates the total emissions of greenhouse gases caused by households and firms in Stockholm.

The lack of discussion of the urban heat island effect noted earlier is another serious drawback. Considering the seemingly rather successful results of city and regional planning since the 1940s, one would have expected a discussion of possibilities to counteract loss of vegetated areas and waste-heat emissions. Since the concentration of greenhouse gases in the atmosphere will go on increasing for many years, irrespective of international and national agreements on reducing current levels of emissions, Stockholm's action plan should also have discussed how Stockholm and the surrounding municipalities can increase their adaptive capacity in order to prevent potential damage that might be caused by recurring heat waves and other kinds of extreme weather, or by rising sea levels.

From a social engineering perspective, it seems even more serious that the plan lacks estimates of the costs for implementing the various measures suggested. To ask for the costs does not imply a downgrading of current environmental threats. It simply reflects the fact that any society's resources are limited; the more resources used for environmental purposes, the less will be available for other important expenditures like schools or care of the elderly. Cost estimates are needed to ensure that a reasonable part of the total budget is used to combat climate change and also to improve the environment in other ways. They are also needed to guarantee an efficient use of this part of the budget. As it is now, it is, for example, impossible to know whether the measures suggested for reducing car traffic would result in as much greenhouse gas reduction per unit spent as those proposed to make housing more energy-efficient. Anyone adopting a social engineering perspective would also be negatively surprised to find no discussion of the need for research and innovations to reduce emissions and improve the environment.

Up till around 1950, Stockholm's growth seems mainly to have been driven by national and international developments. Since then, its role as a national knowledge center has become successively more evident, and growth has partly become self-generating. Furthermore, according to international comparisons of cities and urban regions (see Mälardalsrådet, 2012), Stockholm has recently evolved into one of the leading European metropolitan regions in terms of intellectual capital and its rate of innovation. This should give rise to a more optimistic view concerning the environment. Combining this innovative capacity with the diversity of its industries and political pragmatism, the Stockholm region might, to paraphrase Braudel, become a global "electric transformer," inspiring urban sustainability measures across the world.

NOTES

1   With the exception of Vällingby-Hässelby to the northwest and Sätra-Skärholmen to the southwest, which were added after World War II.
2   Stockholm companies producing telephones, milk separators, drilling equipment, lighthouses, and dynamite.
3   L. M. Ericsson's interaction with the Swedish Telegraph Administration (*Telegrafverket*) exemplifies this national procurement phenomenon, but the importance of a large and trend-oriented local market cannot be overestimated. In 1912, Stockholm had one of the world's highest numbers of telephones per capita: 205 per thousand in Stockholm, 30 in London and 88 in New York (Magnusson, 2010).
4   Great Britain had long been an important trading partner for Sweden; trade with the United States increased primarily after World War I.
5   Several other larger cities were also unincorporated in county councils, as they were considered large enough to handle key county functions such as health care on their own. Stockholm was unique in that it was also subject to its own "Local Administrative Board" – the local arm of the national government – because the Swedish Crown historically preferred to maintain a stronger hold on the capital city than on other municipalities. The municipal government of Stockholm generally considered this unequal treatment to be to its detriment.
6   Or the *Nordic model*; the other Nordic countries have similar approaches and policies.
7   Much of this section is based on the chapters "Den sociala kapitalismen" (Social capitalism) and "Den svenska modellen" (the Swedish model) in Magnusson (2010).
8   A 1965 national government decision to build 1 million rental apartment units over a ten-year period.
9   Editors' note: as was mentioned in Chapter 1, Stockholm's handling of municipal waste is a subject of debate among sustainability researchers. Some applaud Stockholm's investments in efficient combustion and district heating facilities that use municipal waste as a feedstock; others consider the region's relatively high per capita waste production as a liability and consider combustion to be an unsustainable option.
10 Sweden's established international aid target of at least 1 percent of Swedish GDI.
11 In Sweden, Palmstierna (1968) was almost as influential.

REFERENCES

Andersson, Å. E. (1985) *StorStaden, K-samhällets Framtid*, Stockholm: Prisma.
Andersson, Magnus *et al.* (1997) *Stockholms årsringar*, Stockholm: Stockholmia förlag.
Andersson, Monica (2009) *Politik och stadsbyggande. Modernismen och byggnadslagstiftningen*, Stockholm: Stockholm University.
Andersson, O. (2012) *Vykort från Utopia. Maktens Stockholm och medborgarnas stad*, Stockholm: Dokument Press.
Axelsson, K. (2012) *Global miljöpåverkan och lokala fotavtryck*, Stockholm: Stockholm Environment Institute.
Batty, M. (2008) "The size, scale, and shape of cities," *Science*, 319: 769–771.
Börjesson, M. (2012) "Are subway investments profitable?" Stockholm: Expert Group on Public Economics, Ministry of Finance.
Braudel, F. (1985) *Civilization and Capitalism 15th–18th Century*, vol. 1: *The Structures of Everyday Life*, New York: Harper & Row.
Carson, R. (1962) *Silent Spring*, Boston: Houghton Mifflin.
Dufwa, A. (1985) *Stockholms tekniska historia*, vol. 1: *Trafik, broar, tunnelbanor, gator*, Stockholm: Liber förlag.
Eriksson, E. (1990) *Den moderna stadens födelse. Svensk arkitektur 1890–1920*, Stockholm: Ordfront förlag.
Eriksson, E. (2001) *Den moderna staden tar form. Arkitektur och debatt 1910–1935*, Stockholm: Ordfront förlag.
Gullberg, A. (2001) *City: drömmen om ett nytt hjärta. Moderniseringen av det centrala Stockholm 1951–1979*, Stockholm: Stockholmia förlag.
Hall, [Sir] Peter (1998) *Cities in Civilization*, London: Pantheon Books.
Hallerdt, B. (1992) *Stockholms tekniska historia*, vol. 5: *Ljus, kraft, värme*, Uppsala: Almqvist & Wiksell.
Hammarström, I. (1970) *Stockholm i Sveriges ekonomi 1850-1914*, Stockholm: Almqvist & Wiksell.
Hårsman, B. and Quigley, J. M. (2010) "Political and public acceptability of congestion pricing: Ideology and self-interest," *Journal of Policy Analysis and Management*, 29(4): 854–874.
Henderson, J. V. and Thisse, J.-F. (eds.) (2004) *Handbook of Regional and Urban Economics*, vol. 4: *Cities and Geography*, Amsterdam: Elsevier.
Ingo, S. (2002) "Miljöfrågorna i regionplaneringen," in C. Söderbergh (ed.) *Stockholmsregionen 50 år av regionplanering 1952–2002*, Stockholm: Regionplane- och trafikkontoret.
Isaksson, K. (2008) *Stockholmsförsöket: en osannolik historia*, Stockholm: Stockholmia förlag.
Johansson, B. and Strömquist, U. (2002) "Stockholmsregionen: Sveriges tillväxtcentrum." In C. Söderbergh (ed.) *Stockholmsregionen 50 år av regionplanering 1952–2002*, Stockholm: Regionplane- och trafikkontoret.
Johansson, I. (1987) *StorStockholms bebyggelsehistoria. Markpolitik, planering och byggande under sju sekler*, Möklinta, Sweden: Gidlunds.
Kallstenius, P. (2010) *Minne och vision, Stockholms stadsutveckling I dåtid, nutid och framtid*, Stockholm: Max Ström.
Keynes, J. M. (1936) *The General Theory of Employment, Interest and Money*, London: Macmillan.
Larsson, Y. (1977) *Mitt liv i stadshuset*, Stockholm: Stockholms kommunalförvaltning.
Linzie, J. (2002) "Stockholmsregionen i världen." In C. Söderbergh (ed.) *Stockholmsregionen 50 år av regionplanering 1952–2002*, Stockholm: Regionplane- och trafikkontoret.
Lundewall, P. (2006) *Stockholm: Den planerade staden*, Stockholm: Carlsson bokförlag.
Magnusson, L. (2010) *Sveriges ekonomiska historia*, Stockholm: Norstedts.
Mälardalsrådet (2012) *Stockholm Report 2012*, Stockholm.
Ohlsson, P. T. (1994) *100 år av tillväxt*, Stockholm: Brombergs bokförlag.
Palmstierna, H. (1968) *Plundring, svält, förgiftning*, Stockholm: Rabén & Sjögren.
Pettersson, R. (ed.) (2008) *Bekvämlighetsrevolutionen*, Stockholm: Stockholmia förlag.
Sax, U. (1998) *Vällingby: ett levande drama*, Stockholm: Stockholmia förlag.

Schéele, S. (1991) *Stockholmare och andra. Demografiska fakta, utvecklingstendenser och framtidsfrågor*, Uppsala: Konsultförlaget.

Schumpeter, J. A. (1943) *Capitalism, Socialism and Democracy*, London: Allen & Unwin.

Sidenbladh, G. (1985) *Norrmalm förnyat 1951–1981*, Stockholm: Arkitektur.

Snickars, F. and Axelsson, S. (1984) *Om hundra år. Några framtidsbilder av befolkning, samhällsekonomi och välfärd under 2000-talet*, Ds SB 1984: 2, Stockholm: Liber/Allmänna.

Stockholms stads generalplaneberedning (1963) *1962 års cityplan*, Stockholm.

Stockholms stads generalplaneberedning (1965) *Tunnelbaneplan för Stor-Stockholm*, Stockholm.

Stockholms stads stadsplanekontor (1952) *Generalplan för Stockholm 1952*, Stockholm.

Stone, B. Jr. (2012) *The City and the Coming Climate: Climate Changes in Places We Live*, New York: Cambridge University Press.

Strandh, S. (1985) *Från pyramid till laser. Ur teknikens historia*, Stockholm: Natur och Kultur.

Swedish Environment Protection Agency (2012) *Konsumtionsbaserade miljöindikatorer*, Report No. 6483, Stockholm.

Wijkmark, B. (2002) "Regionplaneringen – förspelet; Sex regionplaner; Att utveckla en regiondel," In C. Söderbergh (ed.) *Stockholmsregionen 50 år av regionplanering 1952–2002*, Stockholm: Regionplane- och trafikkontoret.

World Commission on Environment and Development (1987) *Our Common Future* (the Brundtland Report), Oxford: Oxford University Press.

CHAPTER 3
USING THE CONCEPT OF SUSTAINABILITY

INTERPRETATIONS IN ACADEMIA, POLICY, AND PLANNING

Ulrika Gunnarsson-Östling,
Karin Edvardsson Björnberg,
and Göran Finnveden

3.1 INTRODUCTION

ENVIRONMENTAL ISSUES have long been important in Stockholm, although the word *environment* was not used in the political debate before the 1960s (Lilja, 2011). During the 1960s and 1970s, the word gained new ecological meaning: it no longer simply referred to the physical character of the surroundings, but involved a deeper understanding of the interactions among living organisms and between organisms and ecosystems. This semantic shift reflected a change in focus in the City's environmental work. Before the shift, the City was certainly working with issues such as water and air quality, but it did not take a comprehensive approach to environmental issues. In the late 1960s and early 1970s, a new ecological consciousness took shape, among other things making a mark on the City's budget debates. Following the Rio Declaration and *Agenda 21*, the focus of Stockholm's environmental politics and planning changed again, from creating good environmental (ecological) conditions to, in the program accepted in 1995, furthering sustainable development in a wider sense. In recent years, a further shift in focus can be seen, as ecological sustainability is put aside in favour of economic development, or "sustainable growth."

At the city level, the concept of sustainable development is usually put into practice by municipal planners. Although studies exist on how planners and decision makers operationalize the concept of sustainable development into environmental policies and plans (e.g. Porter and Hunt, 2005; Lombardi *et al.*, 2011), important research remains to be done on how the concept is given concrete meaning in everyday planning and decision making. This chapter contributes to this research agenda. The chapter explores how the multidimensional concept of sustainable development, as interpreted and discussed within the academic community and among politicians in Stockholm, affects planning and decision making at the local level. Although there is a "first-level," or surface, understanding of the meaning of sustainable development, both within academia (Baker, 2006; Lafferty and Meadowcroft, 2000) and among politicians in Stockholm, there is little direct or indirect reference to the concept in actual planning. The concept of sustainable development is not explicitly referred to in major plans of relevance for Stockholm, nor does it seem to play any role in the resolution of conflicts between different planning goals. This parallels a depoliticization of environmental issues at the political level in Stockholm. As is described in section 3.3, there is a tendency among politicians to regard environmental problems as something that can be solved through increased knowledge and better technology. This depoliticization of environmental issues "further up" in the administrative chain has implications for local planning. As is described in section 3.4, planners avoid following established sustainability targets, in part because there are no political guidelines on how to handle goal conflicts. Interestingly, they often do so by claiming that it is difficult to define sustainable development.

The chapter is divided into five sections. Section 3.2 provides a brief discussion of the concept of sustainable development. Although it has a widely accepted core meaning, there are many different and sometimes incompatible interpretations of the concept (Connelly, 2007). There are differing views as to the normative principles underlying the concept and how it ought to be operationalized at national and subnational government levels.

Section 3.3 gives a historical overview of how the concept of sustainable development has been operationalized at the political level in the City of Stockholm. It shows how the meaning ascribed to the words *environment* and *sustainable development* has changed over the years. A discursive struggle is going on over what the words should mean. As the chapter shows, there are always multiple discourses present simultaneously; however, one discourse tends to dominate at any given time.

In section 3.4, two planning cases are used as examples of how the concept of sustainable development is operationalized at the planning level in Stockholm: the municipal comprehensive plan and its core concept of "the walkable city," or *Promenadstaden* in Swedish (Stockholm City, 2010), and the plan for national transportation infrastructure (Swedish Road Administration, 2009a). Thus, one plan is national, with local implications, and one is purely local. The two examples show that although formal "requirements" for environmental sustainability exist in political documents adopted both at national and at local levels (goals, indicators, etc.), these requirements are not dealt with explicitly in planning, for example when it comes to choosing between different and conflicting goals.

Section 3.5 provides a discussion of how the discursive issues presented in sections 3.2 and 3.3 are put into context in actual planning and decision making in Stockholm. The depoliticization of environmental issues at the local level, and ensuing unwillingness to explicitly consider value trade-offs, result in a deprioritization of environmental issues within planning processes. One way of mitigating this tendency could be to work with multiple plans, or images of the future, where environmental problems are addressed through different means. By using different alternative futures, planning could become repoliticized and the superficial consensus surrounding sustainable development could be productively challenged.

3.2 THE CONCEPT OF SUSTAINABLE DEVELOPMENT

Sustainable development is a multidimensional concept with many historic roots. It became more widely known to the general public through the Brundtland Report, which defines sustainable development as "development that meets the needs of the present without compromising the ability of future generations to meet their own needs" (World Commission on Environment and Development [WCED], 1987, p. 43). One of the historic roots of the concept of sustainable development was the increasing concern about global environmental degradation and the possible limits to economic growth (Robinson, 2004). Another root was the awareness that developing countries had an urgent need to expand their economies to meet the needs of their people. In the Brundtland Report, the United Nations tries to resolve the conflict between socioeconomic development and protection of the natural resource base through the principle of sustainable development.

Although sustainable development is generally understood as a uniting concept that resolves tensions between social and economic development while also protecting the environment, there is considerable disagreement as to its precise meaning. Although most scholars agree that sustainability is important, diverging views exist as to what actions are required to satisfy human needs, whose needs or interests should take priority when development policies are being designed, and what degree of conflict exists between socioeconomic development and environmental protection. There are many different ways of operationalizing the concept, depending on how sustainability is defined and what issues are perceived to be most important (e.g. Harvey, 1996; Redclift, 2005). As a result, sustainable development, although "the master concept of international discourses of environment and development" (Meadowcroft, 1999, p. 13), has come to be seen by many as "one of the most diversely applied concepts among academics and professionals discussing the future" (Newman and Kenworthy, 1999, p. 1; see also Baker, 2006).

Whose needs should be taken into account?

One of the central controversies surrounding the concept of sustainable development concerns whose needs should be taken into account when designing public policies. The Brundtland Commission defines the concept of sustainability in anthropocentric terms; in other words, the concept is essentially directed towards *human* needs. The needs, or interests, of other living organisms and ecological wholes (e.g. ecosystems and the biosphere) only matter insofar as they contribute to the satisfaction of human needs. Measures to protect the environment are essentially taken because the environment is necessary for human well-being, or *welfare* (Baker, 2006).

Anthropocentric definitions of sustainable development have been criticized in the literature. Some academics dispute the idea that the environment is only instrumentally valuable, valuable only insofar as it contributes to other (instrumental or intrinsic) values. In their view, the environment – including individual organisms, species, and ecosystems – has intrinsic value (e.g. Rolston, 1988). Therefore, development efforts should not primarily be designed so as to guarantee opportunities for human welfare over time, but should also be framed in ways that pay tribute to nature's intrinsic value.

Another point of criticism concerns the lack of action-guidance provided by the concept of sustainable development. Although the concept of sustainable development, as defined by the Brundtland Commission, is straightforward in the sense that it focuses on human welfare, the vagueness of the notion of human needs makes it difficult to operationalize the concept. What needs are we talking about? Are we talking about basic human needs – that is, *needs that must be satisfied* in order to guarantee our immediate survival? Or are we also talking about things we think we need but that are perhaps more adequately described as *things we want*? Obviously, very different policies will have to be adopted depending on what needs (or wants) are taken into consideration.

Furthermore, apart from the most fundamental human needs such as subsistence, protection, and freedom, needs tend to be socially constructed. This means that they often change over time (Redclift, 1992). The needs of people living in ancient Egypt or medieval England were (at least partly) different than the needs of people living today. Even among people belonging to the same generation, needs can differ significantly depending on income level, age, or national or cultural background. This appears to pose a methodological problem: unless we are talking about only fundamental human needs and we agree on what those needs are, how can we use the principle of sustainable development as a normative model for social change? We do not know for sure what future societies will look like or what the needs of future generations will be. How, then, can we design policies based on the idea of non-diminishing opportunities for (human) welfare?

What is needed to satisfy human needs?

In economic literature, meeting people's needs is generally understood as satisfying preferences.[1] Development is sustainable if it satisfies the preferences of the present generation without compromising opportunities to satisfy the preferences of future generations. Thus, Neumayer (2010, p. 7) defines sustainable development as development that "does not decrease the capacity to provide non-declining per capita utility (welfare) for infinity" (see also Rees and Wackernagel, 1996).[2]

Those items that generate the capacity to provide utility are commonly referred to as *capital*. One of the most debated issues relating to the economic concept of sustainable development concerns the question of what forms of capital are needed to provide "non-declining utility for infinity," or the preservation of resources at a level that can meet the needs of coming generations. Two fundamental paradigms exist: weak and strong sustainability (Pearce *et al.*, 1989). Weak sustainability requires that total capital be maintained but allows for substitution among its different parts (natural, human, and man-made capital). This implies that natural capital can safely be diminished so long as it is replaced by an equivalent amount of manufactured capital (Ayres *et al.*, 2001; Munda, 1997). By contrast, strong sustainability assumes that natural capital is fundamentally non-substitutable by other types of capital. It therefore requires that different types of capital be maintained separately. In the most extreme interpretation, strong sustainability requires that all natural resources (species, glaciers, minerals, etc.) be preserved, since they cannot be replaced by anything else (Hansson, 2010). Other interpretations are less extreme and allow for substitution on certain conditions. One such more modest interpretation requires that the aggregate value of the resource stock be kept intact, but allows for substitution between different types of natural capital (Neumayer, 2010). Another more modest interpretation puts restrictions on the substitutability between different forms of natural capital and calls for the preservation of the physical stock of those natural resources that are considered critical for maintaining important and irreplaceable environmental, or life-support, functions (Ekins, 2003; Deutsch *et al.*, 2003).

Arguments have been made for and against both the weak and the strong concept of sustainability. The concept of weak sustainability has been criticized for being redundant in the sense that it does not extend beyond the traditional criterion of welfare maximization (Beckerman, 1994) and for building on an unrealistic assumption, namely that of complete substitutability between different forms of capital. For example, Ekins *et al.* (2003) argue that at least some of the functions performed by natural capital are unique in the sense that they cannot be replaced by other capital components (see also Holling and Meffe, 1996). Basic life-support functions and *amenity services* (e.g., aesthetic, recreational, or scientific value conveyed by a preserved natural environment) are two examples. It is hard to see, it is argued, how manufactured capital could adequately compensate for deteriorations in climate or ecosystem stability, or for aesthetic losses, such as the loss of a landscape or a glacier. The concept of strong sustainability, on the other hand, has been criticized for being unrealistic and impractical in the sense that it demands too much of present generations. Uncertainty about the survival of the human race, it is argued, makes it reasonable to be less worried about the capacity of future generations to satisfy their preferences. The concept of strong sustainability has also been criticized for being morally repugnant, since it appears to divert resources to yet unborn individuals that could instead be used to alleviate suffering of people presently alive (Beckerman, 1994).

What degree of conflict exists between economic growth and environmental protection?

Different sustainability discourses have different views on the degree to which the goals of socioeconomic development and environmental protection conflict. On the one side of the spectrum are traditional neoclassical discourses, which deny that a fundamental conflict exists between economic growth and environmental protection. Environmental degradation occurs because of market failures – that is, the costs of environmental degradation are not properly internalized (not covered for by producers or consumers but left to third parties). If costs were to be internalized, the market would solve the problem of environmental degradation. Hence, in traditional neoclassical terms environmental problems are fundamentally reducible to economic problems, or market imperfections, and policy interventions should be adjusted accordingly.

Other discourses are less extreme but still optimistic toward the possibility of combining economic growth and the protection of the natural capital base. In Sweden, ecological modernization has been the dominant sustainability discourse since the Social Democratic prime minister Göran Persson launched the *Gröna Folkhemmet* (green welfare state) vision in 1996 (Anshelm, 2002, p. 36). The discourse can be operationalized in different ways; however, it typically views continued economic growth as compatible with – and sometimes even necessary for – an environmentally benign development (Hajer, 1995; Hedrén, 2002). Through techno-industrial progress, such as clean technologies, environmental management, and product design, natural resources can be used to further socioeconomic development in a sustainable way.

By contrast, traditional ecological (environmentalist) discourses view nature as something that has to be protected from the negative impacts of economic activity. In this view, socioeconomic development always requires consumption of natural capital in one way or other, which means that technological progress can only go so far in achieving environmental protection. Consequently, additional measures are needed to protect the environment. Historically, environmental protection has been achieved through the establishment of national parks – that is, sanctuaries of wildlife that are cut off and protected from human disturbance. However, preservationist approaches have at times been counterproductive. One prominent example is the Ängsö national park in the Stockholm archipelago, which was established in 1909 to protect the special flora on the island. Severe restrictions were issued concerning grazing on the island, which forced local farmers with grazing cattle to leave. This led to the island quickly being overgrown with shrubs, which posed a significant threat to the meadows and rare species originally intended to be protected by the measures. In this case, the protectionists failed to see that humans played a significant role in the ecosystem they wanted to protect. To see humans only as a disturbing element can thus mean that goals (in this case, the goal of preserving rare plants) are not reached. Instead, it is important to see humans as part of ecosystems, which also implies focusing not only remote natural areas but also on the environment where people spend their everyday lives, which is often in cities.

A more modern ecological discourse suggests that the earth has entered the *Anthropocene*, a geological period in which humanity is the leading agent of environmental change on the planet. In the Anthropocene, ecosystems cannot be seen existing alongside humankind; instead, humans are part of ecosystems and also have a heavy influence on them. Humanity depends on services from ecosystems (Daily, 1997), and it is therefore vital that thresholds that could seriously influence the provision of these services are not passed (Rockström *et al.*, 2009). Management of social-ecological systems becomes important to safeguard the sustainable provision of ecosystem services (Galaz *et al.*, 2012).

What does sustainable development require in terms of intragenerational justice?

Although sustainable development as defined by the Brundtland Commission is framed in terms of justice across generations, it also applies within one generation (intragenerational justice). There are good reasons for extending the principle to the intragenerational case, since it is difficult to see how unequal opportunities for welfare could matter morally in the case of temporally distinct people but not in the case of people belonging to one and the same generation. Accordingly, environmental justice (Harvey, 1996) discourses put intragenerational inequalities at the top of the sustainability agenda.[3] The emphasis within environmental justice discourse on present social inequalities, and the interlinkages between social inequalities and environmental degradation, is supported by studies, primarily from the United States, showing that certain social groups bear a disproportionate burden of environmental problems (Shrader-Frechette, 2002). Thus, in the United

States environmental justice is a politically recognized concept, and the Office of Environmental Justice of the United States Environmental Protection Agency (EPA) makes efforts to integrate environmental justice into all activities performed by the EPA (US EPA, 2012).

Does the proliferation of meanings undermine the usefulness of the concept?

In addition to the differences of opinion that exist regarding the normative foundations of the concept of sustainable development, there are different views on whether the range of meanings present in academic literature poses a problem to actual planning and decision making. Some writers believe that because of its lack of definitional precision, the concept of sustainable development cannot be used as a political reference point (Lélé, 1991; see also Beckerman, 1994; Robinson, 2004); it is nothing but a gun to be hired for any cause. Others, however, regard the vagueness of the concept as "inevitable" (Jacobs, 1999; Connelly, 2007), or even something positive, since it allows people and politicians with different and sometimes conflicting political agendas to agree on common goals (Mitcham, 1995; Robinson, 2004).

Baker (2006) suggests that the complexities of the concept of sustainable development can be captured through the notion of *essentially contested concepts*. Essentially contested concepts have more or less agreed-upon "first-level" political meanings; however, under this apparent unity there is deeper disagreement.[4] The vagueness and open-ended character of the concept of sustainable development makes it easy to modify to fit changing circumstances. It also makes the concept vulnerable to hijacking and redefinition by people and organizations who wish to instantiate the concept to fit their own political agendas.

3.3 THE POLITICAL DISCOURSE ON SUSTAINABLE DEVELOPMENT

From Stockholm City Council's creation in 1863 until today, issues such as air quality and wastewater treatment have been debated by Stockholm politicians, but the word *environment* was not used in the political debate before 1966 (Lilja, 2011). From merely having referred to the physical surroundings, in the late 1960s and at the beginning of the 1970s the term took on an ecological significance and involved the understanding of interactions between living organisms, including humans. There was thus a shift of perspective and an ecological consciousness evolved. Lilja (2011), who has studied the Stockholm's City Council's budget debates between 1961 and 1980, shows that the concept of environment became demanding in its character, which means that all parties were talking about the environment by the end of the 1960s, although there was disagreement about what the term meant. Environmental issues thus became important to many.

In the early 1970s, the environmental concept became more institutionalized in Stockholm, and in 1976 a proposal for Stockholm's first *environmental political program* was launched (the seventh such program was adopted in 2012; it is now simply called an *environmental program*). In a review of these programs, and the political debate about them within the Stockholm City Council, Gunnarsson-Östling (forthcoming) has shown how the environmental discourse that began to emerge in the 1960s changed over time.[5]

The 1976 debate showed a strong consensus among the political parties. Environmental issues were seen not primarily as the result of political decisions, but as something that occurred by accident, and that could be corrected using rational and scientific solutions. These solutions were seen as possible to achieve with more inventories and more knowledge. This is a trend that has persisted, although ideological differences have also been apparent, especially in the 1989, 1995, and 2003 debates, when advantages and disadvantages of capitalist and communist systems were discussed, as was freedom versus control as ways of solving environmental problems. Another dividing line between the parties concerns the distinction between weak and strong sustainability. Over the years, some political parties (above all, the Green Party and the Left Party) have tended to base their arguments on a strong view of sustainable development, where preservation of at least some forms of natural capital is considered necessary. Others (represented mainly by the Moderate Party and the Liberal Party) have usually defended weaker notions of sustainability (Gunnarsson-Östling, forthcoming). To take one example, in the 2003 debate the Liberal Party's Björn Ljung was critical of the environmental program because it "promoted growth neither in the city nor in the region," and instead argued for market-based development:

> If we are to solve the environmental problems in the long term – and we are all agreed on that – it is necessary to see the environment as a part also in growth and development. Then you cannot put up a lot of obstacles. A simple picture is that if you had tried to prevent the growth and development through the years, we would not have received all the good environmental technology that we have now.[6]

The Green Party instead argued that growth does not necessarily equal economic growth:

> Good growth can be located in several areas. It need not just be growth in the economy; it could be growth in quality of life as well. And it need not be growth of automobile traffic; it can also be growth of human capital.[7]

In 1995, a new approach to the environmental program was preceded by a broader consultation with different actors and also the involvement of citizens. This was a direct result of the United Nations' major meeting on environment and development held in Rio de Janeiro in 1992. It highlighted the importance of citizen and other stakeholder participation in decisions and implementation of environmental activities. In the planning of Stockholm, this has, for example, meant that the environmental movement has been given the opportunity to speak up on ongoing plans. However, we know little about the impact that this has had in practice; more research would be needed. This view of democracy is based on a consensus approach and the idea that if only we talk enough with each other, we will probably agree (Dryzek, 2000). Another result from the UN meeting in Rio was that the focus of Stockholm's environmental program from now on was to be geared toward sustainable development rather than just the "protection of the environment."

From 2003 and onwards, Stockholm's environmental program has the stated goal of contributing not only to sustainable development but also to attractiveness (Stockholms stad, 2003). In addition, environmental issues have to a larger degree been pitched as the responsibility of individual citizens and consumers rather than as concerns for collective regulation. Citizens are therefore encouraged to personally shoulder the responsibility of "doing the right thing," through actively choosing to travel by public transport, becoming responsible consumers who consume environmentally friendly products and service, take care of nature, promote biodiversity in the balcony box, sort their garbage, and think about the indoor environment to prevent health problems (Stockholms stad, 2003, pp. 7–8).

In 2008, environmental issues were relativized even further; for example, the program now states that a "healthy environment" may mean different things to different people. The program also states that sustainable development does not stand in contrast to economic growth (Stockholms stad, 2008, p. 3). Even transport, previously framed as the big problem in relation to environmental issues, is now seen as important in the welfare state because "efficient transport is a prerequisite for competitiveness in a global environment" (Stockholms stad, 2008, p. 5, authors' translation). Thus, economic growth has become a compelling concept. In the debate, even the Left Party states that it is excited about economic growth, as increases in revenues will mean that there are more resources to spend on doing good for the environment (Gunnarsson-Östling, forthcoming).

The 2012 program follows the above general line, but with an increased focus on urban development. The economistic and growth-oriented perspective becomes more explicit and the idea of the sustainable city is now further articulated as also being foundationally dependent on technological innovations to solve environmental problems. The program links to the City's *Vision 2030*, which among other things stipulates that

Stockholm in 2030 should have strong international competitiveness and be seen as the main growth region in Europe, and further makes the claim that the city is on track to meet the goal of being fossil-fuel-free by 2050 (Stockholms stad, 2007). Nevertheless, the issue of transport has shifted even more to being a matter of providing efficient transportation as a prerequisite for competitiveness and solving bottlenecks in the region's transportation system (Stockholms stad, 2012, p. 6).

The politicization of environmental issues apparent in the 1989, 1995, and 2003 debates is now gone and instead the debate has become largely a numerical exercise regarding how the targets should be formulated. Again, the environmental issue is thus depoliticized – that is, perceived as something that can be solved efficiently by using more knowledge and science – and the idea of the fundamental necessity of economic growth appears to be wholly unquestioned, along with the status of the car as a crucial transport technology for the society of the future (Gunnarsson-Östling, forthcoming). The emphasis on technological solutions is evident in several of the environmental program's target areas. For example, in the first section, on transport, the program states that "efficient, smooth-running transport systems are essential for competitiveness in a global economy" (p. 6). At the same time, the program acknowledges that road transport "has a negative impact on the city environment in the form of noise, hazardous inner-city air, barriers, and an increased environmental impact" (p. 6). According to the program, this conflict is to be resolved through better technology, such as environmentally certified cars, better infrastructure, and a more cost-efficient use of resources.

The target areas in the latest environmental program are *environmentally efficient transport*, *non-toxic goods and buildings*, *sustainable energy use*, *sustainable use of land and water*, *environmentally efficient waste management*, and *a healthy indoor environment*, and these areas have been fairly intact over time. In these areas, a number of well-defined targets have been formulated. These relate to the national Environmental Quality Objectives.[8] To give an example of the level of detail given, levels of particulate matter (PM10) must meet the norm of 50 _g/m^3, and this is allowed to be exceeded no more than 35 days per year (Stockholms stad, 2012, p. 8). Also, the City of Stockholm has a long-term goal to be fossil-fuel-free by 2050. Should the target be met, greenhouse gas emissions must be reduced by on average 2.5 percent per year, or 10 percent over the program period (2012–2015) (Stockholms stad, 2012, p. 18). However, it is unclear how these targets should be met at the same time that the transportation system, with a large focus on road infrastructure, is being developed.

Interestingly, parts of the program can be interpreted as embracing a moderately strong notion of sustainability. For example, several of the goals in the program's fourth section, on sustainable use of land and water, build on the idea that certain amounts of natural capital ought to be maintained in order to guarantee satisfaction of recreational, educational, aesthetic, or other (human) needs. The program explicitly states that encroachment should be avoided in order to preserve vital (needs-satisfying) functions of ecosystems. Where encroachments into land and water areas do occur, compensation should be paid, not through man-made capital but "by way of an equivalent [ecological] function", or "with an equivalent function for the green qualities of the city" (p. 22) (see targets 4.2 and 4.3).

3.4 WHAT HAPPENS IN REALITY?

Example 1: *The Walkable City*, the City of Stockholm's comprehensive plan

In 2010, the City of Stockholm adopted a new comprehensive plan, called *The Walkable City*, or *Promenadstaden* in Swedish (Stockholm City, 2010). A comprehensive plan is advisory and works as a guiding tool when detailed area plans are developed. The detailed plans are developed for limited parts of the city and are legally binding.

Sustainable development is mentioned explicitly in the plan, which states that "concepts such as sustainable growth and sustainable development are problematic because there are no set definitions and because they contain a number of inherent conflicts" (Stockholm City, 2010, p.10). Then it states that "[t]he Comprehensive Plan is based on Vision 2030, the City of Stockholm's current definition of sustainable growth" (ibid.).

These two quotations are interesting from a number of different perspectives. It is stated that it is difficult to define sustainable development, although there are several official documents that could be used, for example the national environmental quality objectives. Then the terminology partly shifts from *sustainable development* to *sustainable growth*, and then the concept of sustainable growth is given a specific meaning. In this way, the explicit definition of sustainable development is avoided.

The *Vision 2030* document, in turn, does not include any explicit definition of *sustainable growth*, although *growth* is clearly described in the Vision, which states that an increase in population of approximately 20 percent is expected in the coming years. The document also strongly aligns with the ecological modernization discourse mentioned above. It is, for example, stated that "[i]nnovations have resolved many environmental problems" (Stockholms stad, 2007). It also notes that "the transportation system is characterized by environmental technology and logistics that have essentially neutralized carbon emissions" (ibid.). Under the heading "Urban Policies for Sustainable Growth," the plan calls for the Environmental Code to be made more flexible, suggesting that environmental legislation could hinder desired growth and thus constitute an undesirable impediment, which should be avoided.

It is interesting to look at some of the specifics of the plan. In the area of transport, a number of problems are described. They include road congestion, which already exists and is expected to increase, owing to the growing population, emissions of greenhouse gases, which need to decrease significantly, concentrations of air pollutants, which need to decrease, and the need to better connect different parts of the city. The plan then includes suggestions for new transport infrastructure. These suggestions are based on the so-called "Stockholm Agreement" (Stockholmsförhandlingen, 2007). The environmental impact assessment of that agreement concludes that congestion will increase, emissions of air pollutants will increase, and the emissions of greenhouse gases will decrease slightly, but not at all in line with radical decreases that are necessary. It can also be noted that the emissions of greenhouse gases are underestimated – since, for example, emissions from the actual construction of new roads are not included (Finnveden and Åkerman, 2011). Therefore, the consequences of the comprehensive plan are not in line with the intentions of the plan. This is, however, not commented on in the plan. Furthermore, the plan would not satisfy the conditions of *sustainable development* as interpreted in line with national policy documents such as the national environmental quality objectives – although, by avoiding the use of the concept "sustainable development," this situation is avoided.

The statement made in *Vision 2030* that by 2030 the transportation system will have essentially neutralized carbon emissions is not at all supported by the environmental impact assessment. Instead, the environmental impact assessment predicts significant emissions of gases contributing to climate change. The reasons behind the lack of coherence between *Vision 2030* and the environmental impact assessment of the Stockholm Agreement are not known to us. The *Vision 2030* document does not give any clues as to why it anticipates much more radical changes in transport emissions than those detailed within the environmental impact assessment, which is charged with describing the likely development of such factors.

The Walkable City puts an emphasis on green spaces. It considers them to be important, but also states that conflicts may exist between the preservation of green spaces and urban development. In the plan, the focus is on the direct use of green spaces. For example, the title of the part of the plan dealing with green spaces is "Sport, Recreation and Attractive Green Spaces." Indirect uses of green spaces, such as for air purification, noise reduction, climate regulation, and other ecosystem services, are not discussed. The more modern ecological discourse mentioned above is thus not present in the plan. If there has to be a weighting between the importance of green spaces and urban development, the plan's non-recognition of the indirect uses of green spaces could influence that balance.

Further, the discourse of environmental justice is not present in *The Walkable City*, although the plan will influence the distribution of environmental "goods" and "bads." The transportation plans in the Stockholm

Agreement are expected to lead to reduced emissions in the affluent center of the city and increased emissions in the less well-to-do suburbs. This result can be compared to the development of income distribution in Stockholm, which is similar: while incomes are increasing in the City of Stockholm, much slower increases are seen in the suburbs.

Example 2: Transportation infrastructure

This example concerns the plan for national transportation infrastructure, which, although formulated at the national level, has implications for Stockholm, since many of the investment objects in the plan are connected to the Stockholm region. The plan was developed by Swedish transport agencies[9] and submitted to the national government (Swedish Road Administration, 2009a). An environmental assessment of the plan was also submitted (Swedish Road Administration, 2009b). The plan concerns the period 2010–2021 and includes suggestions for new investments and maintenance of the Swedish transportation system corresponding to approximately €50 billion. The focus is on roads and railroads.

One investment proposed in the plan is the Stockholm Bypass project, an orbital highway link located at the western edge of the City of Stockholm that is expected to cost more than 25 billion Swedish kronor (approximately €2.9 billion), making it the most expensive road project in the plan. The Stockholm Bypass is also part of the *Vision 2030* document and the Stockholm Agreement. It is, however, quite controversial. The bypass is a motorway that will run mostly in tunnels west of Stockholm and connect the northern and southern parts of the Stockholm region. Advocates of the motorway argue that it is cost-efficient and important for a growing city. Opponents point to the increased trips that the road will generate and suggest that investments in public transport could fulfill the same function and contribute to a more sustainable transport system.

The overall objective of Sweden's national transport policy is to ensure the economically efficient and sustainable provision of transport services for people and businesses throughout the country. More specifically, the transport sector should contribute to achieving the environmental quality objective "Reduced Climate Impact," and other environmental quality objectives for which the development of the transport system plays an important role. The objective "Reduced Climate Impact" requires significant reductions in greenhouse gas emissions. To be in line with the two-degree target for climate change, the transport sector needs to reduce greenhouse gas emissions by 40 percent by 2020 compared to 1990 levels, 80 percent by 2030, and 95 percent by 2050 (Swedish Road Administration, 2009c).

The plan states that economic efficiency, measured using cost–benefit analyses, has guided its development (Swedish Road Administration, 2009a). No corresponding comment is made with regard to sustainable transport services. One reason for this discrepancy may be the existence of established methods for evaluating economic efficiency; according to the plan, corresponding tools to quantify sustainability are lacking, and sustainability was therefore not evaluated.

The environmental assessment concludes that the national infrastructure plan:

- will lead to increased negative impacts on biological diversity (which is in direct conflict with the national environmental quality objective, "A Rich Diversity of Plant and Animal Life");
- will contribute only in a limited way to the achievement of the environmental quality objective "Clean Air";
- will not decrease the number of people affected by noise above the reference values decided by the Parliament, thus failing to contribute to sustainable development with regard to human health and a good environment.

In relation to greenhouse gas emissions, the plan claims to lead to small emission reductions (Swedish Road Administration, 2009a, b). It is thus clear that the planned projects (referenced in Swedish Road Administration, 2009c) do not contribute to the significant decreases in emissions required to reach climate goals. Furthermore, the agencies have underestimated energy use and greenhouse gas emissions in several ways, for example by failing to include emissions from the actual construction of infrastructure included in the plan (Finnveden and Åkerman, 2011). It is therefore likely that the plan will rather lead to increased greenhouse gas emissions.

It is interesting to note that despite the environmental assessment's use of the term *sustainable development*, the plan itself largely avoids discussing sustainable development, sustainable transportation, or similar terms. The plan does not include definitions of these terms, or evaluations of its own sustainability. In this way, the conclusion that the plan does not support sustainable development is avoided (Finnveden and Åkerman, 2011).

3.5 DISCUSSION AND CONCLUSIONS

A number of observations can be made concerning how the concept of sustainable development is interpreted and used in academia, as well as in concrete policy and planning documents in Stockholm, and Sweden generally. First, there is often broad agreement among politicians regarding the general structure of the concept of sustainable development. One illustration of this is the different lists of sustainability indicators that have been developed by the United Nations and other international and national organizations. These lists often have similar content, indicating a broad con-sensus on what aspects should be considered when operationalizing the concept of sustainable development. A broad consensus can also often be seen in Swedish policy making. One example is the 16 National Environmental Quality Objectives operationalizing the ecological dimension of sustainable development, which have been approved by the Swedish Parliament in broad consensus. Similarly, within Stockholm City Council, politicians largely agree on the importance of environmental issues and about which environmental goals and targets should be adopted at the local level. However, underlying this agreed-upon "first-level," or surface, understanding of what sustainable development is, there is considerable disagreement about the means that should be employed to further sustainable develop-ment and on the trade-offs that have to be made among different goals. For example, the goal of reduced climate impact may be interpreted either as being achievable through a reduction in car traffic, or as the natural con-sequence of environmentally superior automotive technologies.

Second, as described in section 3.3 it could be argued that there exists a tendency among politicians in Stockholm to depoliticize environ-mental matters and to (over)emphasize the value of political consensus. In the City's view, environmental problems can and should be addressed using improved knowledge and more efficient technologies. The emphasis of the latest Stockholm Environment Program (2012–2015) on technological solutions to conflicts between transportation and environmental protection provides one example of this approach.

Third, with regard to planning, two observations can be made. Section 3.4 shows that the concept of sustainable development is largely ignored within planning processes. There are also conflicts between what is stated in the plans and the environmental assessments that are produced. For example, *Vision 2030* suggests that the transportation system will be essentially carbon-neutral. This is in conflict with the environmental assess-ments of both the Stockholm Agreement and the plan for national trans-portation infrastructure, which show that the emissions of greenhouse gases will not decrease significantly. In *Vision 2030* and the comprehensive plan, *The Walkable City*, this conflict between what is stated in the planning document and the conflicting result of the environmental assessment of an important part of the plan (the Stockholm Agreement) is avoided by simply not undertaking an environmental assessment of the comprehensive plan. Making it compulsory to conduct environmental assessments of compre-hensive plans could possibly make this conflict more apparent.

In addition, section 3.4 shows how the tendency to depoliticize environmental issues at the local level – encapsulated in the idea that "if we gain better knowledge about means, this alone will help us reach the goals of sustainable development," and the resulting unwillingness to explicitly consider (politically laden) value trade-offs – has implications for local planning. Planners in Stockholm have difficulty working toward established goals of sustainable development, since there are no political guidelines on how to handle goal conflicts. In practice, planners therefore avoid following established sustainability targets. They do this by claiming that it is difficult to define sustainable development, thus ignoring the possibilities that exist. In this way, they avoid drawing the conclusion that the plans are in conflict with sustainable development, which of course is important, since it would be more difficult to implement a plan that explicitly supports unsustainable development.

Framing sustainable development as something apolitical may sound like a favor to planners, who are supposed to work in a non-political way. However, this results in planners suggesting only one plan, one solution, that neither defines the concept of sustainable development in a precise way nor highlights conflicts with other societal goals. Thus, planning documents are permeated by a superficial consensus. We instead suggest that several plans, or images of the future, could be worked out, clearly building on different interpretations of which goals are important for sustainable development. This would mean that the development of Stockholm is described through, say, both the goal of reaching the environmental goals that have been agreed on, and other societal goals. However, in order to avoid a singular focus on technological issues, it is also important to ask how people would like to live their lives. Since people have different desires and moral values, there will always be conflicts about what constitutes a good or sustainable city. Planning could productively highlight these conflicts, instead of pretending that there is consensus.

NOTES

1   As has already been noted, it could be questioned whether satisfying people's preferences amounts to meeting their needs. It could be argued that sometimes what a person really needs is for his or her present preferences not to be satisfied. Hausman and McPherson (2006, p. 120f.) mention the case of a youngster wanting a powerful motorbike. Satisfying this preference, no matter how strong it is, is not necessarily what the youngster needs.

2   It is the *capacity* to provide non-declining per capita utility for infinity that counts; non-declining utility itself is not required (Neumayer, 2010; see also Howarth, 1997; Page, 1983).

3   Environmental justice is an anthropocentric discourse. However, there are writers who do not focus solely on humanity's access to environmental qualities or vulnerability to environmental risks. Low and Gleeson (1997), for example, understand *justice within the environment* as the distribution of environmental values within human populations, while *justice to the environment* is about how humans treat non-human nature.

4   The concept of justice is a good example of an essentially contested concept. Most people agree that justice is about giving people what is due to them. However, many different views exist as to what it means to give people what is due to them (Swift, 2006, pp. 11–12).

5   A discourse is usually described as a way to talk about, think about, and interpret phenomena such as the environment and sustainable development (see, for example, Dryzek, 1997; Winther Jørgensen and Phillips, 2000). This approach means that the way to talk about the environment and sustainable development is not seen as clearly reflecting the outside world, but it also has an active part in creating the world because some actions become relevant and others unthinkable (Hansen and Simonsen, 2004; Winther Jørgensen and Phillips, 2000).

6   Minutes of Proceedings for Stockholm City Council meeting, held at Stockholm City Hall, Stockholm on Monday 17 February 2003. Statement by Björn Ljung, Liberal Party, p. 51 (authors' translation).

7   Minutes of Proceedings for Stockholm City Council meeting, held at Stockholm City Hall, Stockholm on 17 February 2003. Statement by Viviann Gunnarsson, Green Party, p. 52 (authors' translation).

8   The national Environmental Quality Objectives are a system of 16 environmental quality objectives and many more interim targets first adopted by the Swedish government in the late 1990s (Swedish Ministry of the Environment, 2001; Swedish Ministry of Sustainable Development, 2006). The objectives describe the quality and state of the environment regarded as sustainable in the long term, with the intention of providing a coherent framework for environmental programs at national, regional, and local level.

9   Editors' note: Sweden had separate transportation agencies for road, rail, sea and air transport until 2010, when a common transportation administration comprising several transport modes was created.

REFERENCES

Anshelm, J. (2002) "Det gröna folkhemmet. Striden om den ekologiska moderniseringen av Sverige," in J. Hedrén (Ed.) *Naturen som brytpunkt. Om miljöfrågans mystifieringar, konflikter och motsägelser,* Stockholm: Symposion.

Ayres, R. U., van den Bergh, J. C. J. M., and Gowdy, J. M. (2001) "Strong versus weak sustainability: Economics, natural sciences, and 'consilience'," *Environmental Ethics*, 23(2): 155–168.

Baker, S. (2006) *Sustainable development,* London: Routledge.

Beckerman, W. (1994) "'Sustainable development': Is it a useful concept?" *Environmental Values*, 3: 191–209.

Connelly, S. (2007) "Mapping sustainable development as a contested concept," *Local Environment*, 12(3): 259–278.

Daily, G. C. (ed.) (1997) *Nature's Services: Societal Dependence on Natural Ecosystems,* Washington, DC: Island Press.

Deutsch, L., Folke, C., and Skånberg, K. (2003) 'The critical natural capital of ecosystem performance as insurance for human well-being," *Ecological Economics*, 44(2–3): 205–217.

Dryzek, J. S. (1997) *The Politics of the Earth: Environmental Discourses,* Oxford: Oxford University Press.

Dryzek, J. S. (2000) *Deliberative Democracy and Beyond: Liberals, Critics, Contestation,* Oxford: Oxford University Press.

Ekins, P. (2003) "Identifying critical natural capital: Conclusions about critical natural capital," *Ecological Economics*, 44(2–3): 277–292.

Ekins, P., Simon, S., Deutsch, L., Folke, C., and De Groot, R. (2003) "A framework for the practical application of the concepts of critical natural capital and strong sustainability," *Ecological Economics*, 44(2–3): 165–185.

Finnveden, G. and Åkerman, J. (2011) "Not planning a sustainable transport system: Swedish case studies," in *Proceedings of World Renewable Energy Congress 2011,* 8–13 May, Linköping, Sweden.

Galaz, V., Biermann, F., Folke, C., Nilsson, M., and Olsson, P. (2012) "Global environmental governance and planetary boundaries: An introduction," *Ecological Economics*, 81: 1–3.

Gunnarsson-Östling, U. (forthcoming 2013) "Mellan ekologi och tillväxt. Miljöpolitiska handlingsprogram i Stockholm 1976–2012" (Between ecology and growth: environmental action plans in Stockholm 1976–2012), in T. Nilsson (ed.) *Du sköna nya stad. Privatisering, miljö och EU i Stockholmspolitiken* (Brave new city: Privatization, environment and EU in Stockholm politics), Stockholm: Stockholmia förlag.

Hajer, M. A. (1995) *The Politics of Environmental Discourse: Ecological Modernization and the Policy Process*, Oxford: Oxford University Press.

Hansen, F. and Simonsen, K. (2004) *Geografiens videnskabsteori: En introducerende diskussion. Frederiksberg*, Roskilde, Denmark: Roskilde Universitetsforlag.

Hansson, S. O. (2010) "Technology and the notion of sustainability," *Technology in Society*, 32: 274–279.

Harvey, D. (1996) *Justice, Nature and the Geography of Difference*, Malden, MA: Blackwell.

Hausman, D. M. and McPherson, M. S. (2006) *Economic Analysis, Moral Philosophy, and Public Policy*, Second edition, New York: Cambridge University Press.

Hedrén, J. (2002) "Naturen som hot mot det moderna: några ideologkritiska reflektioner," in J. Hedrén (ed.) *Naturen som brytpunkt: Om miljöfrågans mystifieringar, konflikter och motsägelser*, Stockholm: Symposion, pp. 298–333.

Holling, C. S. and Meffe, G. K. (1996) "Command and control and the pathology of natural resource management," *Conservation Biology*, 10, 328–337.

Howarth, R. B. (1997) "Sustainability as opportunity," *Land Economics*, 73(4): 569–579.

Jacobs, M. (1999) "Sustainable development as a contested concept," in A. Dobson (ed.) *Fairness and Futurity: Essays on Environmental Sustainability and Social Justice*, Oxford: Oxford University Press, pp. 21–45.

Lafferty, W. M. and Meadowcroft, J. (2000) "Introduction," in W. M. Lafferty and J. Meadowcroft (eds.) *Implementing Sustainable Development: Strategies and Initiatives in High Consumption Societies*, Oxford: Oxford University Press, pp. 1–22.

Lélé, S. M. (1991) "Sustainable development: A critical review," *World Development*, 19(6): 607–621.

Lilja, S. (2011) "'Miljö' som makt i Stockholmspolitiken 1961–1980," in T. Nilsson (ed.) *Stockholm blir välfärdsstad: Kommunpolitik i huvudstaden efter 1945*, Stockholm: Stockholmia förlag, pp. 295–331.

Lombardi, D. R., Porter, L., Barber, A., and Rogers, C. D. F. (2011) "Conceptualising sustainability in UK urban regeneration: A discursive formation," *Urban Studies*, 48(2): 273–296.

Low, N. P. and Gleeson, B. J. (1997) "Justice in and to the environment: Ethical uncertainties and political practices," *Environment and Planning A*, 29: 21–42.

Meadowcroft, J. (1999) "Planning for sustainable development: What can be learned from the critics?" in M. Kenny and J. Meadowcroft (eds.) *Planning Sustainability: Implications of Sustainability for Public Planning Policy*, London: Routledge, pp. 12–38.

Minutes of Proceedings for Stockholm City Council meeting, held at Stockholm City Hall, Stockholm on 17 February 2003, Stockholms kommunfullmäktiges handlingar 2003, Protokoll 2003:2.

Mitcham, C. (1995) "The concept of sustainable development: Its origins and ambivalence," *Technology in Society*, 17(3): 311–326.

Munda, G. (1997) "Environmental economics, ecological economics, and the concept of sustainable development," *Environmental Values*, 6(2): 213–233.

Neumayer, E. (2010) *Weak versus Strong Sustainability: Exploring the Limits of Two Opposing Paradigms*, 3rd ed., Cheltenham, UK: Edward Elgar.

Newman, P. and Kenworthy, J. (1999) *Sustainability and Cities: Overcoming Automobile Dependence*, Washington, DC: Island Press.

Page, T. (1983) "Intergenerational justice as opportunity," in D. MacLean and P. G. Brown (eds.) *Energy and the Future*, Totowa, NJ: Rowman & Littlefield, pp. 38–58.

Pearce, D. W., Markandya, A., and Barbier, E. (1989) *Blueprint for a Green Economy*, London: Earthscan.

Porter, L. and Hunt, D. (2005) "Birmingham's Eastside story: making steps towards sustainability?" *Local Environment*, 10(5), 525–542.

Redclift, M. (1992) "The meaning of sustainable development," *Geoforum*, 23(3): 395–403.

Redclift, M. (2005) "Sustainable development (1987–2005): An oxymoron comes of age," *Sustainable Development*, 13(4): 212–227.

Rees, W. E. and Wackernagel, M. (1996) "Urban ecological footprints: Why cities cannot be sustainable – and why they are a key to sustainability," *Environmental Impact Assessment Review*, 16(4–6): 223–248.

Robinson, J. (2004) "Squaring the circle? Some thoughts on the idea of sustainable development," *Ecological Economics*, 48, 369–384.

Rockström, J., Steffen, W., Noone, K., Persson, Å., Chapin, F. S. III, Lambin, E. F., Lenton, T. M., Scheffer, M., Folke, C., Schellnhuber, H. J., Nykvist, B., de Wit, C. A., Hughes, T., van der Leeuw, S., Rodhe, H., Sörlin, S., Snyder, P. K., Costanza, R., Svedin, U., Falkenmark, M., Karlberg. L., Corell, R. W., Fabry, V. J., Hansen, J., Walker, B., Liverman, D., Richardson, K., Crutzen, P. and Foley, J. A. (2009) "A safe operating space for humanity," *Nature*, 461: 472–475.

Rolston, H. III (1988) *Environmental Ethics: Duties to and Values in the Natural World*, Philadelphia: Temple University Press.

Shrader-Frechette, K. S. (2002) *Environmental Justice: Creating Equality, Reclaiming Democracy*, New York: Oxford University Press.

Stockholm City (2010) *The Walkable City: Stockholm City Plan*, Stockholm: City Planning Administration.

Stockholms stad (2003) *Stockholms Miljöprogram 2002–2006: På väg mot en hållbar utveckling*, Stockholm.

Stockholms stad (2007) *Vår vision*. Retrieved from www.stockholm.se/OmStockholm/Vision-2030/Innovativ-och-vaxande/

Stockholms stad (2008) *Stockholms Miljöprogram 2008–2011. Övergripande mål och riktlinjer*, Stockholm.

Stockholms stad (2012) *Stockholms miljöprogram 2012–2015*, Stockholm.

Stockholmsförhandlingen (2007) *Samlad trafiklösning Stockholmsregionen för miljö och tillväxt – till 2020 med utblick till 2030*, Överenskommelse mellan Stockholms-regionen och staten December 2007. Retrieved from www.regeringen.se/content/1/c6/09/47/70/f3df2c98.pdf (accessed 3 May 2013).

Swedish Ministry of Sustainable Development (2006) "Environmental quality objectives: A shared responsibility," Summary of Government Bill 2004/05:150, Stockholm.

Swedish Ministry of the Environment (2001) "The Swedish environmental objectives: Interim targets and action strategies," Summary of Government Bill 2000/01:130, Stockholm.

Swedish Road Administration (2009a) *Nationell plan för transportsystemet 2010–2021*, Vägverket.

Swedish Road Administration (2009b) *Miljökonsekvensbeskrivning för Nationell plan för transportsystemet 2010–2021*, Vägverket.

Swedish Road Administration (2009c) "Vägverkets handlingsplan för begränsad klimatpåverkan." Publikation 2009:82, Vägverket.

Swift, A. (2006) *Political Philosophy: A Beginner's Guide for Students and Politicians*, Second edition, Cambridge: Polity Press.

US Environmental Protection Agency (2012) *Environmental Justice*. Retrieved from www.epa.gov/Compliance/environmentaljustice/ (accessed 29 June 2012).

Winther Jørgensen, M. and Phillips, L. (2000) *Diskursanalys som teori och metod*, Lund, Sweden: Studentlitteratur.

World Commission on Environment and Development (1987) *Our Common Future*, Report of the United Nations World Commission on Environment and Development, Oxford: Oxford University Press.

CHAPTER 4
A SUSTAINABLE URBAN FABRIC

THE DEVELOPMENT AND APPLICATION OF ANALYTICAL URBAN DESIGN THEORY

Lars Marcus, Berit Balfors, and Tigran Haas

4.1 INTRODUCTION

STOCKHOLM HAS A HISTORY of claiming international leadership in urban planning and design. In the 1950s, Vällingby rose to world renown as a Swedish interpretation of the neighborhood unit concept; recently, Hammarby Sjöstad has taken on a similar role in sustainable urban design. Vällingby was a successful response to the particular needs of the industrial society and its concentrations of skilled but not very specialized labor, which in the Swedish context was often socially homogeneous and implied lifelong employment with large companies or public organizations, with generally small pay differentials. For this, the extended neighborhood unit model that Vällingby represented, with a wider scope of functions, including workplaces and commercial activities, in a city district otherwise dominated by housing, presented a humane environment of modest but high-quality urban design, housing, and, not least, extensive green areas. The low density, large number of green areas, and high degree of zoning naturally generated a great transport demand that was taken care of by what was at the time a world-class public transport system – which, we might note, was developed not for environmental but for economic reasons, since it was calculated that most people at that time were not able to afford a private car. What is easily forgotten today is the ethnic, social, and to a high degree even economic homogeneity of Swedish society at the time – a homogeneity that meant that most of the many neighborhood units built in Stockholm after World War II (of which Vällingby was the most ambitious) generally had a similar identity, something that over time was to change.

To a surprising degree, Hammarby Sjöstad can be said to build on the Vällingby model, in that the coordinated planning of transit and urban development that began in the 1950s set Stockholm on a path toward becoming a sustainable transit metropolis; regional mobility and a focus on preserving nature created the prerequisites for what was to come. At the same time, Hammarby Sjöstad is clearly something different, a response to an environmentally alert knowledge society, where the need for high concentrations of labor is still essential, but now it is labor of a much more diversified and specialized kind that is needed. As such, a greater need for knowledge exchange and meetings in person, at least in the professional segments of the labor market, is implied. The Vällingby model does not typically offer such spaces of exchange; this is why Hammarby and many other more recently built areas of Stockholm are characterized by higher densities, shorter distances (because there is far less green space), and public places designed for intense use. Goals favoring high density have primacy; equally ambitious sustainability targets are primarily achieved through the application of various sorts of technology. Consequently, compared to the earlier suburban landscape of neighborhood units in an extensive green structure, what appeared in the early 1990s was a new, specifically Swedish, interpretation of ecological urban planning and design based on green, sustainable, and compact development ideals – for which Hammarby was to prove the flagship.

In a wider and more theoretical perspective, however, projects such as Vällingby and Hammarby reflect a long history of normative ideas about how to best build cities to achieve better societies, often based on speculative thinking by architects. It was this tradition that, according to Françoise Choay (1997), was formalized in "theories in urbanism" in the late nineteenth century. While *speculative* here clearly should not imply any pejorative reading (and neither should *architects*), it is necessary to acknowledge that such theory constitutes a particular type of knowledge that cannot be accountable beyond its limits. Such theories are what Herbert Simon (1969) called theories of possibility – of how things could or ought to be – rather than theories about actuality – about how things are. Given current global challenges (not least in urban development) raised by the multifaceted crises we are witnessing, a practice that is only based on such theory is unsatisfactory.

In this chapter, an epistemological discussion of the knowledge foundations of urban design, not least sustainable urban design, is attempted as a background to a presentation of both some of the current research directions in this field at KTH and of the strong connections between such knowledge foundations and current urban design practice in the City of Stockholm. We argue that the new challenges currently facing the field call for something of a knowledge revolution, where urban designers need to move from being experience-based craftspeople to theory-based professionals. On the one hand, this implies greater knowledge about ecological systems in cities. On the other hand, and more pertinent to the field, it requires a more systematic understanding of the prime material – spatial form – through which one can structure and shape different urban systems, including ecological systems. In section 4.2, we review current theories and trends in sustainable urban development. In section 4.3, we present different examples of the knowledge needed from the ecological and environmental fields. In section 4.4, we present a concept of more systematic knowledge about spatial form. Finally, in section 4.5 we discuss a real case in Stockholm where such combined knowledge about ecological systems and spatial form was applied. The chapter concludes with a discussion of design theory. This discussion tries to clarify why, today more than ever, there exists a need for an urban design theory that is to a much greater degree founded on analytical knowledge, and is complementary to the already strong speculative strands in the field.

4.2 CONTEMPORARY TRENDS IN URBAN DESIGN

In an international context, the theories and ideals dominating today's urban design discourse have been examined and defined in various ways, resulting in differing categorizations and definitions such as "territories of urban design," "images of perfection," "urban design force fields," "integrated paradigms in urbanism," "urbanist cultures and approaches to city-making," "new directions in planning theory," "models of good design," and "typologies of urban design." However, the three dominant ideals that stand out are *New Urbanism*, *posturbanism*, and *sustainable urbanism* (Haas, 2008, 2012), with other important categories – such as *everyday urbanism*, *ecological urbanism*, and *landscape urbanism* – qualifying as runners-up.

In the end, all share a concern with shaping and composing public spaces; and creating livable and healthy places of variation, interest, familiarity, interaction, and contrast. They do, however, differ in approach. Some work in the traditional way, advocating public squares and perimeter blocks as integral to shaping and composing cities, based on historical and proven principles (New Urbanism and sustainable urbanism); while others turn to vanguard approaches referring to globalization, mediatization, and contemporary architectural transformation (posturbanism, city marketing, and place branding schemes). Some look for narratives and hidden dimensions in the micro-sphere of the public realm (everyday urbanism), and still others turn to solutions connected with urban ecological design and systemic landscape design schemes (ecological and landscape urbanism).

Most often writing themselves into the long tradition described above, these approaches all share a bias toward speculative theory, wherein the analytical support for the solutions proposed is often elusive. This places considerable stress on practices in the field, which as a result lack knowledge support when facing the major new knowledge challenges that arise from global demographic and consumption changes, growing social inequalities, peak oil, global warming, and biodiversity loss (Brito and Stafford-Smith, 2012). As the world is becoming more urbanized and the majority of people live in cities and urbanized and suburbanized regions, cities need to be on the front line in dealing with these urgent issues. We also see how these changes are starting to make an impact, altering the requisites for future planning, urban design, and the role of the professional as we know it. In the face of this, it is disconcerting that knowledge in the field of urban design, which structures and shapes the urban fabric at the scale at which people actually perceive and experience cities, relies so heavily on theory that is weak in its analytical foundations. That is not to say that more research is the only solution – scientific knowledge is truly as risky to apply indiscriminately as speculative knowledge – but we clearly see a task for research to contribute to stronger analytical foundations in urban design theory. Steps towards this are also currently being taken (Droege, 2007; Beatley *et al.*, 2009; Palazzo and Steiner, 2011). In the following subsection, we take a closer look at the theories we find most influential in developing sustainability in urban design and, more specifically, address how they have been received and interpreted in a Swedish context.

Major strands in sustainable urban design

Sustainable urbanism is a phrase that is used widely, often with ecological and green connotations, constituting a rather complete framework for the interdisciplinary planning and urban design of contemporary cities, neighborhoods, and settlements. It explores sustainability and urban design in a holistic manner by focusing on the processes that shape the form and function of our built environment in its full complexity – the infrastructures, land developments, built landscapes, social networks, systems of governance and economics, and facilities that all collectively make up metropolitan regions (Calthorpe, 2010; Farr, 2007; Haas, 2012). In practice, sustainable urbanism focuses on identifying small-scale catalytic interventions that can be applied to urbanized locations – interventions that, in aggregate, aim to achieve an overall shift toward sustainable neighborhoods, districts, and regions. In its fullest meaning, sustainable urban design refers to those ways in which the agendas of cities and of urbanism, and those of environment, conservation, and sustainability, can and do profoundly overlap.

The approach includes the following elements: building and growing more densely and compactly; integrating transportation means, patterns, and land use; creating walkable mixed-use urban environments that permit and encourage walking and cycling through the implementation of car-free areas; investments in public transit and transportation; creating closed-loop urban eco-metabolisms and self-sustaining agricultural systems with local production of foods, goods, and materials; creating health and environmental benefits by linking humans to nature, including walk-to open spaces; neighborhood storm-water systems, waste treatment, and food production (permaculture); and investment in and commitment to sustainable, renewable, and passive technologies integrated into the built form (e.g. solar, wind, and biomass).

In parallel, a slightly different paradigm has emerged: *ecological* and/or *landscape urbanism*. In contrast to New Urbanism or posturbanism, architecture and urban design are not the main foci here. The view of ecological urbanism is that the fragility of the planet and its resources present an opportunity for speculative design innovations rather than technical legitimation for promoting conventional solutions. This approach also imagines an urbanism that has the capacity to incorporate and accommodate the inherently conflicting conditions of ecology and urbanism (Corner, 1999; Mostafavi *et al.*, 2010). Instead of relying solely on buildings as the medium of design, landscape urbanism uses landscapes: infrastructure, public space, and open space – an approach that is much more comfortable with open-endedness. Landscape urbanism really plays two instruments, one being environmental science and the other being design culture. As such, it constitutes a combination of eco-knowledge and urban planning and design (Waldheim, 2006). Ecological and/or landscape urbanism is an actual city-making approach achieved by a strong ecological, urban design, and landscape fusions, the realization of and adaptation to the complexity presented by climate change, and resilience and energy-saving schemes as foci.

Reception and application in the Swedish context

These theories, especially sustainable urbanism, have been (and are being) interpreted and "glocalized" in urban design terms in the Swedish context. This new approach is geared more toward an urban ecological design, mostly in terms of energy efficiency and less in terms of creating complete neighborhoods through complexity and completeness. Issues of compactness/density and connected tissues in terms of traffic patterns have been more or less resolved, but the element of *place-making* – the creation of convivial neighborhoods with affordability, accessibility, and availability – still has a long way to go. Further, the nexus of New Urbanism's *neighborhood planning*, which can be located within the sustainable urbanism paradigm, has been neither accepted nor transferred to Sweden, at least not in its entirety.

Notwithstanding this, Sweden has advanced its own, rather original, approach to sustainable urbanism. As it did in earlier times, Sweden (and, in particular, Stockholm) has aspired to new leadership here. The large and ambitious brownfield development project Hammarby Sjöstad has found international renown as a landmark case in sustainable urban design. As in the case of Vällingby, what impresses visitors to Hammarby is more than its architectural accomplishments; it is the municipal planning coordination of different urban systems and their responsible administrations and the way in which these systems and administrations work together toward a common goal (Figure 4.1, p. 79). Once an industrial wasteland, good urban design and careful sustainable planning have transformed this area of urban blight into a recognized model for eco-minded change. All buildings have been designed to meet the goals of reducing waste, specifically in terms of CO_2 emissions. Roofs (some of them green) conceal solar panels; handrails also hold solar strips to fuel energy needs, and waste treatment uses some of the most advanced possible techniques, for example a vacuum system. To further enhance traditional methods of recycling and reuse, the district has its own pilot sewage treatment center. The urban design concept is linked to ecological thinking in terms of reforming water and sewage technology, recycling, keeping environmentally sound materials in mind, and heating buildings with renewable fuels. It is, however, weaker when it comes to a deeper integration of local ecosystems and the understanding of the landscape propagated by ecological and/or landscape urbanism.

Residents are drawn to the area because the very buildings themselves embody their values. Additionally, and in line with sustainable modal choices in Stockholm, Hammarby Sjöstad has a diverse system of transportation to serve its residents. The light-rail infrastructure has been developed with four stops in the heart of Hammarby, and it connects directly to the subway network in Stockholm. There are also plans to extend the tram further eastward to connect directly to one of Stockholm's main transport hubs. The area, furthermore, has a successful car-sharing pool. Hammarby also provides its residents with preschool and elementary school facilities, cultural activities, children's spaces, and other amenities for residents that are intended to invite people into the public realm. In terms of urban design, aside from the Swedish legacy of planning and urban design the development of the district was inspired by, and in some respects utilized, principles of New Urbanism, Transit Oriented Development (TOD) and Smart Growth (Congress for New Urbanism, 1999; Haas, 2008; Dittmar and Ohland, 2003; Duany *et al.*, 2009) as well as ecological and/or landscape urbanism, through measures such as sensitive treatments and the planting of shores.

BOX 4.1: HAMMARBY SJÖSTAD SUSTAINABLE CITY

Hammarby Sjöstad is the starting point for the Sustainable City concept and an inspiration for future projects, as well as the foundation of the international branding concept SymbioCity. The project was produced by the City of Stockholm together with several real estate developers. The Sustainable City concept was developed by Sweco, a leading Swedish consultancy providing services in the fields of urban planning, architecture, engineering, and environmental technology, and Professor Ulf Ranhagen at KTH/Royal Institute of Technology, School of Architecture and the Built Environment.

This concept is a working method for sustainability reviews based on open, creative, and constructive communication and cooperation between decision makers, experts, and the public. Important steps include diagnosis of the current situation, specification of key issues and objectives, impact analyses, and selection of strategies. A number of tools are utilized to

Source: Lennart Johansson, InfoBild.

facilitate this process, such as SWOT analysis, scenario techniques, back-casting, and strategic environmental assessment (SEA). In a Sustainable City, the use of renewable resources is emphasized, resource consumption is minimized, and resources are managed in a way that maximizes recovery and reuse. New system solutions provide scope for synergies between sewage, waste, and energy production, and enable coordination with efficient land use, landscape planning and transport systems. This is illustrated by the eco-cycle model, which is essential for a definitive shift from linear to circular resource flows.

The concept is applicable in the planning of new cities and towns, where there is a "window of opportunity" to reduce energy demand by up to 75 percent and achieve an energy supply based on renewables. In addition, the concept can be used to develop strategies for successive realignment of existing urban areas in a sustainable direction. The results are genuinely successful in Hammarby: 40 percent less environmental stress, 50 percent less eutrophication, 45 percent less ground-level ozone, and 40 percent less water consumption.

(Source: SymbioCity, Stockholm City)

SymbioCity: the concept behind Hammarby Sjöstad

The close relation between the Hammarby Model as a new urban design and integrated systems approach (Figure 4.1) and research at the Department of Urban Planning and Environment of KTH (the Royal Institute of Technology) (Ranhagen and Groth, 2012) is important to the perspective explored in this chapter. The urban strategy that provided the foundation for Hammarby Sjöstad was later refined in the *SymbioCity* model. Through the promotion of the model, Hammarby Sjöstad thus became a showcase for a large-scale approach that was designed and developed by Swedish experts at the request of the Swedish government and the Swedish Trade Council. The model demonstrates the importance of coordinating urban functions and the advantages of a holistic approach to urban planning – moves that can result in reduced energy consumption, enhanced accessibility, increased social integration, and a better development of values. By applying a holistic approach to the improvement of the complex urban environment, it creates a model that takes into account the governance aspects coupled with urban planning and urban design, as well as the management of various urban subsystems. Taken together, these form the basis for an integrated and interdisciplinary approach to urban development through a systems approach to waste, energy, water, and sewage, as well as public transport, culture, and nature. The concept has been applied in countries as diverse as China, South Africa, India, and Canada. Within Europe, it has been applied for instance in Narbonne in France, Cork South Docklands in Ireland, and the London

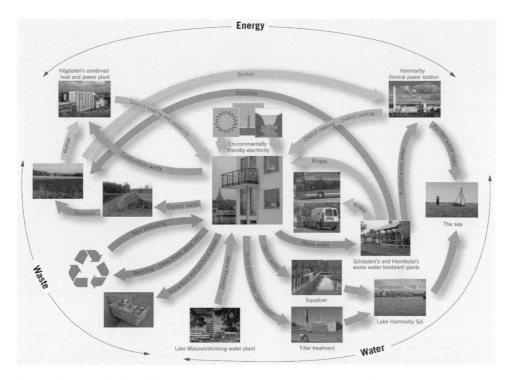

Figure 4.1 Diagram of the SymbioCity concept, an integrated systems approach to sustainable development that is the basis for the Hammarby Sjöstad project and in critical parts developed by the consultancy SWECO and Ulf Ranhagen, KTH School of Architecture and the Built Environment

Adapted from Ranhagen & Groth, 2012

Olympic Village in the United Kingdom, although its first application was in Hammarby Sjöstad.

The SymbioCity approach focuses on how urban governance, urban planning, education, IT concepts, public participation, and other coordinating activities can be used to promote sustainable urbanization. Functioning institutions are crucial in this respect, as well as the coordination of public and private sectors. The SymbioCity concept is thoroughly systemic in its foundations, where all the parts that make up a sustainable city play a crucial and equal role, thereby linking different sectors, actors, and systems. The results are fully optimized and create synergies between the systems in the city: energy, waste management, water supply and sanitation, traffic and transport, landscape planning, sustainable architecture, and urban functions. These sectors typically live their own lives independent of one another, leading to sub-optimization. The SymbioCity approach finds links between sectors and their system investments in order to optimize results (Figure 4.1). All these integrated approaches try to take advantage of synergies between the urban systems of the city, supporting a vision of an up-to-date, people-focused, climate-neutral, and environmentally sustainable society, while at the same time acknowledging the many trade-offs inherent in urban planning. As Rutherford (2008) points out, the Swedish (Stockholm) version of the unitary (sustainable) networked city was supported to a large extent by the decentralization of many of the responsibilities and mandates associated with urban and welfare policies through their transfer to the municipalities.

While the SymbioCity concept and the area of Hammarby Sjöstad can be considered success stories, there has also been clear criticism. This has, on the one hand, concerned social issues: Hammarby Sjöstad to a far greater degree than intended has become an area for an urbanite elite who might share the ethos of the area but simultaneously live a high-consumption lifestyle. This paradox has been the focus of much attention, not least in the daily press. This is expressed, for instance, in reports that the number of private cars in the area has far surpassed planners' intentions. There has also been critique concerning the ambitions of the Swedish government to market and export this planning model, especially to China. This critique has focused on the ethical as well as technological incommensurabilities that follow such an export – for instance, solutions have at times more or less been directly projected from a Swedish context to a Chinese one without adaptation. This critique is related to the worldwide trend of city branding, where sustainability often plays a central role in Swedish cases, but where the aims clearly are not only altruistic but form part of a highly competitive strategic program. The marketing-led perspective has, moreover, made it important for Stockholm to develop a sequel to Hammarby Sjöstad that can continue to promote the city as a leader in environmentally friendly development. On the basis of knowledge and experiences from Hammarby Sjöstad, Stockholm has consequently launched a new project area development, *Norra Djurgårdsstaden* (Stockholm Royal Seaport). This development aims to be more advanced with respect to sustainability, through the implementation of increased environmental requirements for buildings, technical installations, and traffic solutions. One example of the mechanisms to be deployed lies in the recycling systems for water and waste (other examples are described in Figure 4.1).

BOX 4.2: THE STOCKHOLM ROYAL SEAPORT

The vision for the Stockholm Royal Seaport is to use its proximity to water and to the unique Royal National City Park that surrounds the urban development area. The City of Stockholm promotes Stockholm Royal Seaport as "a global showcase for sustainable urban construction and design, where innovative environmental technology and creative solutions are developed, tested and presented." The Royal National City Park is a large area protected by a special law in order to preserve its nature and culture for future generations, and when inaugurated was the world's first ever urban national park.

To link the Stockholm Royal Seaport and the National City Park, landscape elements such as oak trees have been highlighted as especially valuable and important to protect. In addition, endeavors will be made to plant new oak trees in the development area. The development is located in a brownfield area that has been used for various industries, such as gas works and harbor activities, which raises the need for soil remediation. According to Stockholm City, the former gas works (also referred to in Chapter 2) will be "transformed into a vibrant environment with art galleries, open stages and other culture attractions." The existing oil shipping harbor will be relocated to other parts of the Stockholm region, while the ferry and cruise ship terminal will be developed and integrated into the future city district.

The Stockholm Royal Seaport benefits from the environmental experiences drawn from Hammarby Sjöstad but goes a step further, making the new development area a unique climate *positive* urban development project in order to demonstrate that cities can reduce carbon emissions and grow in climate-friendly ways. In order to achieve compliance with the environmental profile of the development urban area, the agreements with the developers contain detailed sustainability requirements regarding housing construction. Examples of sustainability requirements for the Stockholm Royal Seaport include environmental quality certifications, climate change

Source: City of Stockholm and BSK Architects.

adaptation, outdoor environment, energy system, recycling systems for water and waste, environmental friendly transportations, and construction and materials.

An important aspect in the development of the Stockholm Royal Seaport is the close cooperation between public authorities, the developers, and the business and industry throughout the planning and building process. This entails a commitment among the main actors involved to work toward achieving the sustainability objectives for the urban district. Several research projects are connected to the development of the Stockholm Royal Seaport related to energy (e.g. smart grids, energy-efficient construction), ecology (e.g. biodiversity), and procurement.

(Source: City of Stockholm)

4.3 ANALYTICAL THEORIES OF SUSTAINABLE CITIES

A new role for green areas in cities

Thus, we see an extensive range of new approaches to urban planning and design in response to the environmental crises, but we can also conclude that so far the majority of the projects that have been realized and the strategies that are institutionalized, including the SymbioCity approach, are, generally speaking, found within the sustainable urbanism paradigm. In recent years, however, one can identify a growing interest and a developing knowledge base in research and practice, which to a higher degree has its foundations in the ecological urbanism paradigm. Maybe we can call this a shift from a first generation of research and practice in sustainable urban development, primarily addressing climate change, to a second generation that broadens the field to encompass ecological systems in general and biodiversity in particular.

An expression of this lies in the way in which green areas in cities are starting to play an increasingly important role, far beyond the recreational uses traditionally assigned to them. Their planning, design, and maintenance as ecological systems in urban areas rapidly is moving to the forefront. Earlier urban design has seldom seen its task as being one of supporting eco-systems in the city, and opportunities for doing so have often been missed. Naturally, such a task would also imply a greater understanding of the functionality of urban ecosystems; this is why it is often landscape architects who currently take the lead here. By extension, this also means that a far deeper understanding of urban ecosystems and the way in which they interlock with human systems is needed, which makes it necessary to move beyond the kind of speculative theory discussed earlier and develop support in genuine analytical theory from necessary disciplines. While such attempts are to be found within both ecological and landscape urbanism, and to a certain extent also within sustainable urbanism, including the SymbioCity concept, they are still limited and do not take in the full range of necessary knowledge that has developed in recent decades.

Urban ecosystem services

Ecosystem goods (such as food) and services (such as waste assimilation) are benefits that human populations derive, directly or indirectly, from eco-system functions (Costanza *et al.*, 1997). The term *ecosystem services* has evolved from a long lineage of awareness of human dependence on the environment (Wilkinson, 2012). In the Millennium Ecosystem Assessment (2005), ecosystem services are grouped into four main categories: pro-visioning, regulating, supporting, and cultural. In cities, green areas are of primary interest in their role as producers of ecosystem services – that is, services that ecosystems produce to the benefit of humans often without any costs. These include micro-climate regulation, air filtration, and rainwater drainage (e.g. Bolund and Hunhammar, 1999). Another example of an ecosystem service related to green areas is pollination. Ahrné's (2008) study on bumblebees and urbanization found that when one is considering how to support a relatively high number of bumblebee species, the importance of areas with high diversity (such as allotment gardens) needs to be taken into consideration. The study concludes that there is a need to actively plan the larger urban landscape for the benefit of these species. Such a position also includes consideration of qualities related to biodiversity, which refers to the variety of living organisms in urban areas. Biodiversity is essential for the functioning and sustainability of an ecosystem. Different species play specific roles and have specific functions, and changes in species richness and functional type affect the efficiency with which resources are processed within an ecosystem (Alberti, 2005). Biodiversity contributes to the sustain-ability of the ecosystem and, by extension, the services that it generates (Elmqvist *et al.*, 2003). To promote biodiversity, green areas should offer suitable living conditions (habitat) to meet the specific needs of various species, such as certain types of vegetation. Generally, fragmented green structures, typical for urban areas with less connectivity, often have difficulty in offering such conditions, while the larger connected areas with natural and semi-natural vegetation that can be found in peri-urban districts in Stockholm are essential for maintaining biodiversity in the whole region. Within the central districts, on the other hand, green areas are often smaller but can have high ecological qualities. Connecting these areas with the green areas in the peri-urban districts can facilitate the migration of species as well as supporting ecosystem services in central areas (see examples of case studies in Mörtberg *et al.*, 2012; Ernstson *et al.*, 2010). Hence, green areas in urban and peri-urban areas are important for biodiversity. Given the right planning and design, green areas also accommodate a variety of social benefits such as recreation, education, and health.

In recent decades, urban green areas have also received growing attention in relation to climate change adaptation. To mitigate the impacts of extreme weather events in urban regions, such as heat waves and heavy rainfall, green areas are important for the urban climate and storm-water retention (Gill *et al.*, 2007). In addition, green areas can contribute to the removal of significant amounts of air pollution (Shan *et al.*, 2007). Because of the intense competition for land normally found in urban contexts, green areas are not only to be considered sustainable when they possess high natural values; in order for them to offer ecosystem services efficiently, green areas should be supported with institutional and technical systems for management and development.

These ideas are closely related to the concept of urban metabolism, which analyzes cities by drawing analogies with the metabolic processes of organisms. Thus, urban metabolism involves the flows and storage of energy, water, and other materials, the physical transport associated with these flows, and the institutional, economic, and technical systems that manage flows (Kennedy *et al.*, 2007). Cities should be considered to be dependent on both their regional and their global surroundings. The concept of urban metabolism demonstrates the need to address institutional and societal contexts when discussing the conservation, maintenance, and development of green areas.

Reconciling conflicts between human and ecological systems

At the same time, there clearly are conflicts between societal (human) and ecological systems in cities. For instance, movement in ecological systems such as green corridors easily comes into conflict with movement in (human) societal systems, such as traffic arteries, and certain species are not welcomed in cities for health reasons. Yet many perceived conflicts reflect outdated paradigms in our thinking. For instance, the idea that urban and agriculture functions are incompatible is a twentieth-century conception with little support in earlier history. On the contrary, history is replete with examples of urban agriculture throughout the twentieth century even in the West. This has also been the case in Stockholm, where, for instance, allotment gardening has been practiced since the late nineteenth century in very central parts of the city. According to a study by Björkman (2012), interest in allotment gardening has increased since the beginning of 2000. Further, a number of municipalities in Sweden are developing guidelines for urban gardening in order to meet the demands of new users such as schoolchildren and immigrants. Still, some conflicts truly need introspective consideration and knowledge development, not least when it comes to informing urban planning and design.

As a result of ongoing urban development in the Stockholm region, the environment in urban areas is affected both by new land-use claims in existing green areas and by additional emissions of noise, air, and water pollutants that impact an already strained living environment (Balfors *et al.*, 2005, 2010). As a result of the continuous urbanization process, green areas are transformed into areas for housing, industry, and infrastructure. As a consequence, the living conditions for flora and fauna in these and related areas change due to processes of fragmentation and habitat loss, affecting biodiversity in these areas (Balfors and Mörtberg, 2006; Mörtberg *et al.*, 2006). To sustain or develop biodiversity in an urbanized region, a number of interrelated ecological conditions have to be fulfilled. One is the existence of ecological networks that consist of core areas, corridors, and buffer zones. Other preconditions are connectivity to facilitate the exchange between core areas, and time continuity regarding natural habitats (Zetterberg *et al.*, 2010). In the Stockholm region, several studies have investigated the eco-logical conditions that support biodiversity in urban landscapes in transition (Mörtberg *et al.*, 2012; Zetterberg, 2011; Andersson and Bodin, 2009; Borgström *et al.*, 2006). Planning decisions that lead to changes in the com-position and pattern of the landscape impact biodiversity. The impacts may occur on-site, affecting merely the area of development, or off-site, since ecological processes such as species persistence and dispersal often work at large scales. Therefore, a site-based approach is not sufficient for the consi-deration of biodiversity in impact assessment. Instead, it is necessary to consider the quality, quantity, and spatial cohesion of natural habitats and the persistence requirements of species and communities in the entire landscape.

A landscape approach to biodiversity assessments is also consis-tent with Swedish government targets that state the need for methods for assessing the impacts of human actions on biodiversity at a landscape level. Such methods allow for the analysis of the cumulative impacts of many single planning decisions, which may lead to substantial alterations, including the loss, isolation, and disturbance of natural habitats. Many of the processes involved have a temporal and spatial dimension, and can be quantified, analyzed, and visualized with geographical information systems (GIS). When these processes are quantified, they can be used in GIS-based habitat modeling in order to study and predict effects of changes on biodiversity components (Gontier *et al.*, 2010). Spatial predictions can be made in relation to large areas, and the long-term effects of planning scenarios can also be addressed. GIS-based habitat modeling has been used by the City of Stockholm within different planning activities, and by the Stockholm County Administrative Board for the development of a regional management plan for the Eurasian lynx in the county (Zetterberg, 2011).

Predictive tools in urban planning and the design of ecosystems

The integration of biodiversity issues in the assessment of urban develop-ment plans and projects requires prediction tools that employ relevant knowledge about the impact of land-use changes on the fauna and flora inhabiting the area (Gontier *et al.*, 2006). These tools can allow an assessment of the ecological impacts at a landscape level, primarily based on landscape-ecological knowledge. In addition, the tools should allow for prediction of changes in vegetation and the effects of development scenarios at different time scales. A wide range of GIS-based ecological models can be used as prediction tools for biodiversity assessments in, for example,

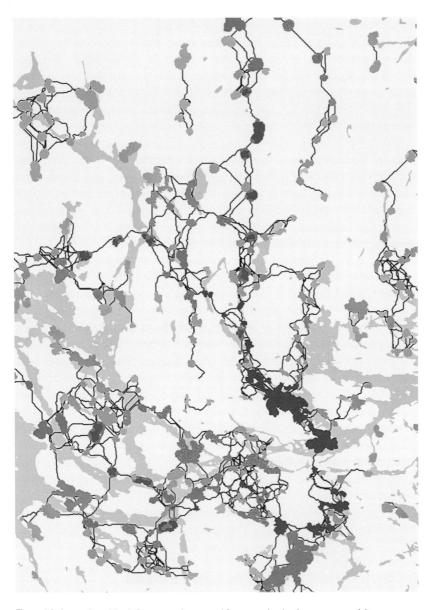

Figure 4.2 Output from MatrixGreen, a software tool for assessing landscape connectivity

EU-regulated decision support processes such as environmental impact assessment (EIA) and strategic environmental assessment (SEA). Selection of the most appropriate model depends on the aim and scope of the study, and the context in which the result will be used. Further issues that need to be considered are, for instance, which biodiversity components are to be modeled, the availability and quality of data and expert knowledge, the time frame, available resources, and the competence of those carrying out the analyses.

Recently, Zetterberg (2011) presented a toolbox with network-based landscape-ecological methods and graph-theoretical indicators, which can be effectively implemented by practitioners within physical planning and design to analyze landscape connectivity. MatrixGreen was developed by Bodin and Zetterberg (2010) and is now freely available at www.matrix green.org. This tool is used by consultants, municipal ecological experts, and other stakeholders involved in planning and decision making.

Biodiversity considerations in environmental impact assessment and strategic environmental assessment

Prior to a decision on a project such as the construction of a road or a new housing area, Swedish legislation requires that an environmental impact assessment (EIA) be carried out, in which the impacts of the project are identified. However, initial decisions on urban expansions and major infrastructure investments are often made at a strategic stage where the long-term development of the region is determined in policies, plans, and programs. For a number of these types of decisions such as the approval of municipal comprehensive plans, a strategic environmental assessment (SEA) is therefore required, addressing the environmental impacts of strategic decisions. The main rationale for applying an SEA is to help create a better environment through informed and sustainable decision making (Fisher, 2003). In addition, SEA contributes to a more effective promotion of sustainable development and allows for the assessment of cumulative impacts (Glasson *et al.*, 1999; Therivel, 2010). However, the high level of abstraction evidenced in policies, plans, and programs involves major methodological problems for the prediction of impacts (Hildén *et al.*, 1998).

The enactment of EU Directive 2004/35/EC, concerning the assessment of the effects of certain plans and programs on the environment, strengthens the need for environmental considerations to be taken into account in physical planning. In this way, nature conservation and biodiversity issues should be integrated in the early stages of the planning process when planning for urban development and infrastructure investments. The Convention on Biodiversity also emphasizes the need to take biodiversity issues into account in environmental assessments. Efforts therefore need to be made to reinforce biodiversity considerations in EIA and SEA.

In order to protect green areas in urban regions such as Stockholm, the impacts of plans, projects, and activities that may contribute to fragmentation and habitat loss need to be considered in planning and decision making. A consistent assessment of potential impacts can strengthen biodiversity considerations in the planning process and contribute to the preservation of biodiversity in the long term. This is also in line with national and international political ambitions, which require the protection of green areas so as to maintain biodiversity in urban regions. In addition, analytically more informed urban design can enhance new development projects that can improve ecosystem services' performance in the Stockholm region.

Existing ecological networks will be affected as a result of habitat loss and fragmentation. Hence, planning should have access to knowledge on ecosystem dynamics at the landscape level, in order to preserve nature and biodiversity values, or at least mitigate the impacts of ongoing developments on these values, which in the long term would contribute to sustainable urban development. In order to counteract a further degradation of the urban environment, the role of disciplinary expertise regarding, for example, the impact of human activities on ecological processes therefore has to be strengthened in planning and decision making.

4.4 TOWARD AN ANALYTICALLY SUPPORTED URBAN DESIGN THEORY

Urban design that includes ecological systems

A second generation of practice and research in sustainable urban development comprising both a deeper and a wider understanding of ecological systems in cities calls for stronger support from analytic knowledge of the kind discussed in the preceding section. This has quite new implications for urban design and, by extension, research in urban morphology, which we here define as the spatial structure generated by the physically built fabric of cities. The first generation, as illustrated by sustainable urbanism and to some extent New Urbanism, admittedly stresses the integration of more environmentally advanced technological systems that can enhance the performance of the urban fabric when it comes to sustainability issues such as energy and waste disposal systems and (more conspicuously) public transport systems. However, these still remain within the limits of rather conventional urban design practices, albeit technologically enhanced. The second generation, on the other hand, asks for a more direct treatment of urban form: how can urban form harbor not only social and economic systems, which it has always done, but ecological systems as well? As was mentioned earlier, new perspectives on the city–nature dichotomy open our eyes to the fact that ecosystems have always played a fundamental role in urban development, for example by supporting urban agriculture. For this reason, it is becoming clear that the strict distinction between urban and natural landscapes is a twentieth-century myth. By extension, this asks new questions of research, also on quite deep theoretical levels, since if we are to develop knowledge about how urban form can harbor not only social systems but also ecological ones, we need to bridge the knowledge divide between human and ecological systems.

In recent years, the KTH School of Architecture has, in collaboration with the Royal Swedish Academy of the Sciences and Stockholm University, been involved in the development of exactly such an alternative approach by aiming to establish a new research field called *spatial morphology of social-ecological urban systems* (Marcus and Colding, 2011), which integrates resilience theory and recent developments in spatial morphology and design theory. This is coupled to the idea of a professional practice called *social-ecological urban design* (Barthel *et al.*, 2013), which can be interpreted as an analytically enhanced development of ecological and/or landscape urbanism. The approach draws, on the one hand, on resilience theory (originally developed by the Canadian ecologist C. S. Holling, 1973), and on the other on spatial morphology as developed in space syntax research (Hillier and Hanson, 1984). Resilience theory developed as a way to understand non-linear dynamics in natural systems, such as the processes by which ecosystems maintain themselves in face of natural disturbance. Most importantly, natural systems are here seen as integrated with human systems in what is called *social-ecological systems*, bridging the human–environment dichotomy (Folke, 2006). The attempt to integrate natural systems into planning in a way that extends beyond a purely protectionist approach is particularly important; the concept of ecosystem services again proves to be critical in confronting this task by making explicit the tremendous number of services that ecosystems perform for the benefit of human systems (e.g. pollination, air and water cleaning, natural drainage). The aim is then to translate such knowledge about resilience, social-ecological systems, and ecosystem services into spatial form, thereby making it applicable in urban design practice. For this, we need an advanced theory of spatial form that addresses the urban micro scale that is particular to urban design. This is what, in recent decades, has developed in the research program *space syntax*, which – interestingly – has its roots in the spatial discipline of architecture rather than geography, as is most often the case for such theories.

Space syntax: an analytical approach to urban morphology
Space syntax research has taken the lead in the development in recent decades of more analytical approaches to spatial morphology. While spatial modeling of cities recently has seen rapid development, not least as a result of the development of GIS, it has almost been completely concerned with

analysis of aggregated levels of urban systems (Batty, 2005). This is in part due to the difficulties of gathering fine-scale data, but more importantly it can be traced to the difficulties in constructing relevant models of urban space at a finer scale. There is therefore agreement that knowledge regarding how to model what can be called *the cognitive level of urban space* – the level at which people on the street experience the city – is underdeveloped. Many of these issues are directly addressed in space syntax, where an analytical approach is taken to the cognitive level of urban space – that is, to the scale where urban space is perceived by individuals at street level – with the specific aim of developing knowledge that can support architectural and urban design (Hillier, 1996; Hillier and Hanson, 1984). Sustaining the system perspective of spatial modeling, but adding more imaginative geometric description from urban morphology, space syntax merges both traditions into an analytical urban morphology that focuses on space. Hence, space syntax can be interpreted as an original and successful school within a wider and to a large extent undeveloped research field that might suitably be called *spatial morphology*.

Based on space syntax modeling, a long series of studies from around the world have found strong correlations between the configurative properties of urban form and pedestrian movement (Hillier *et al.*, 1993; Hillier and Iida, 2005). Moreover, a large body of research (e.g. Greene *et al.*, 2012; Koch *et al.*, 2009) has found correlations between urban form and other urban phenomena, where movement works as the intermediary. For example, it is not surprising to find that streets that are highly accessible in the street system and collect a lot of movement also become prominent locations for retail. This has been confirmed by many studies (e.g. Scoppa, 2012; Bernow and Ståhle, 2011). Taking this a step further, it also seems likely that such streets may in the long run gather higher rents for the letting of floor space; in fact, this has been also confirmed (e.g. Desyllas, 2000; Netzell, 2010). Other studies show similar correlations between urban form and urban phenomena such as crime (Hillier and Xu, 2004), social segregation (Legeby, 2010; Vaughan, 2007), and perceived accessibility to green spaces (Ståhle, 2005).

There is an obvious connection here to the MatrixGreen modeling software, which also builds on graph-theoretical descriptions of spatial networks, the difference being that space syntax clearly deals with human systems, hence the stress of cognitive descriptions, while MatrixGreen deals with ecological systems and landscape connectivity. Clearly, an important strand for further research, currently being developed in collaboration between Stockholm University and KTH, is to develop geometric descriptions of the movement of different species groups to investigate to what degree such spatial patterns could be combined with human systems. By extension, this implies an investigation of the degree to which humans and other species could share the same urban spaces, something that could inform and support urban planning and design in the development of social-ecological urban design.

Spatial capital: toward an analytical theory in spatial morphology

Space syntax research has broken new ground and opened up the challenging cognitive level of urban space for analysis and knowledge development. This is important to the approach to sustainable urban development discussed in this chapter for two reasons. First, for research to further develop a bottom-up understanding of cities as systems of complexity (that is, systems of non-linear relations between variables, where changes on lower spatial scales can have dramatic emergent effects on higher scales), it is strategic to model and analyze the cognitive level of urban space (Batty, 2005). Second, for professional practice to develop empirically informed interventions in urban systems through urban design (where spatial form in the end is the prime medium), new knowledge on the variables of spatial form at work at the detailed, cognitive level is required. Such analytical theory has also been successfully introduced into professional consultancy practice.

At the KTH School of Architecture, research expanding space syntax research into a wider field of spatial morphology has been undertaken over a ten-year period. This work has proceeded along three lines. First, there is a strand developing analytical tools that incorporate geographical data into the purely geometric analyses characteristic of space syntax (Ståhle *et al.*, 2005), wherein a decisive step is the development of the GIS-based *Place Syntax Tool* (Figure 4.3). Second, there is a strand developing new geometric descriptions and measures that aim to capture spatial variables other than those normally included in space syntax research, such as spatial density and diversity (Marcus, 2010a). This has, among other things, generated the concept of *compact sprawl*, whereby measures of density and measures of open space are integrated into a conceptual framework aiming to overcome the dichotomy between densification and the preservation of green areas in urban design (Ståhle, 2008). Third, there have been a range of strategic empirical studies that critically discuss the spatial variable in other fields of research addressing urban systems, such as *social segregation* in urban sociology (Legeby, 2010), local markets in economic geography (Bernow and Ståhle, 2011), the planning of urban green areas in landscape architecture (Ståhle, 2005), densification in urban design (Ståhle, 2008), knowledge environments in innovation systems theory (Koch and Steen, 2012), and walkability in medicine and health studies (Choi, 2012).

These methodological developments and empirical studies are currently synthesized within a theoretical framework called *spatial capital* (Marcus, 2010a), which develops an architectural model of the city based on cognitive descriptions of space. The model is based on a theoretical conception of cities as spatial networks, wherein different locations and their

Figure 4.3 Image created using the Place Syntax Tool, a GIS-based software tool for calculating typical distance measures such as closeness and betweenness in the built fabric at the scale of individual street segments with the possibility of weighing the analysis with geographical data (in this case, population). It was developed by Alexander Ståhle, Lars Marcus, and Daniel Koch, of the KTH School of Architecture, and Anders Karlström, KTH division for Transport and Location Analysis (Ståhle *et al.*, 2005). The image shows a betweenness analysis for the complete pedestrian street and path system in Stockholm municipality, with a radius of 2,000 meters (network in black and white), and accessible population from each and every address point in the system, following the same street and path network within a radius of 500 meters (dots in color in the original) (Legeby, 2010).

interrelations are given particular use-values, and by extension exchange values, depending on their positions in the network, and therefore can be said to constitute spatial capital. The model contributes to theoretical explanations of the social, economic, and ecological systems pertaining to the spatial dimension of cities, but also informs professional practice about how urban systems can be created, developed, and maintained through urban planning and design.

4.5 ALBANO: TOWARD PRINCIPLES OF SOCIAL-ECOLOGICAL URBAN DESIGN

When researchers from the fields of architecture, urban morphology, and systems ecology participated in a real-life development of a new university campus in Stockholm called Albano, the knowledge base in spatial morphology was combined with resilience theory and research on social-ecological systems in urban landscapes (Colding *et al.*, 2006; Andersson *et al.*, 2007; Barthel *et al.*, 2010a; Ernstson et al., 2010). The project was instrumental in the development of the approach to sustainable urban design we call *social-ecological urban design* (Barthel *et al.*, 2013).

This collaboration was supported by the public real estate company Akademiska hus, and was later adapted by the City of Stockholm in its formal plans for the area. The project concerns a new university campus in Stockholm at the strategic site of Albano, located at the crossroads between the city's three major universities, with substantial potential to support a greater exchange between the faculties at the different universities through greater proximity and shared amenities. The site is also located at the border between the inner city (with all its urban attractions) and the Royal National City Park (with its unique natural and historic qualities). Thus, the aim has been to develop an urban campus that ties the established universities together and at the same time contributes to and enhances the natural and ecological qualities in the area through the concept of social-ecological urban design.

As a support for the design of the area, three major urban form elements were identified, namely g*reen arteries*, *active ground*, and *performative buildings*, which bear some similarity both to established concepts within urban morphology (streets, plots, and buildings) and to notions of connectivity, heterogeneity, and density in urban ecology (Alberti, 2008). However, these elements were analyzed using more formalized interpretations developed in the spatial capital concept: distance, density, and diversity (Marcus 2010a; Marcus and Colding, 2011). While the latter variables are normally linked to social and economic processes that are of relevance to urban design, in this project they were developed also to cover relevant ecological processes in cities.

Just as any social urban process requires accessibility between different nodes of activity, ecological processes are also in need of accessibility between nodes of activity, such as habitats. Hence, the notion of the "street" and other similar spaces of connection could be augmented to also include ecological connectivity zones. In this project, such spaces were called *green arteries*, and while they call for design informed by ecological and not only social needs, on a generic level they have the potential to support ecological systems as well as social systems. In a similar vein, cities have throughout their history been characterized by the division of land into discrete plots for different activities and/or owners. Such land subdivision has proven to have positive effects on the development of diversity, in terms of both social and ecological systems in urban areas (Colding, 2007; Marcus, 2000). This was addressed in the study called *active ground*. This element of land subdivision, it was stressed, could be broadened into a tool for the development and support not only of diversity in economic activity or ownership, but of biodiversity as well. Finally, just as we can see buildings as a manifestation of the need to locally enhance and intensify land use through built form, buildings could be extended to support both social and ecological needs, for example through the design of green roofs, green walls, social-ecologically integrated water and air systems, etc. In the project, these were described as aspects of *performative buildings*, implying that they were performative not only in a technical and social sense but also in an ecological sense – that is, the buildings formed intensified nodes in a spatial system that supported not only social systems, as in any urban project, but the local ecosystem as well. Finally, there is the need to design these elements into a systemic whole, a task for which the kind of analytic tools presented here such as the Place Syntax Tool and MatrixGreen could prove critical. Taken together, the development of knowledge about such urban design elements and their systemic relations could make urban design a way to support and enhance local and regional ecological systems rather than pose a threat to them. In turn, urban development of this kind might overcome the city–nature dichotomy.

In conclusion, the project emphasized that the three elements of urban form (green arteries, active ground, and performative buildings) need to be embedded in different institutional settings such as property rights, social networks, and local cultures, maintained and managed over time. Altogether, knowledge on these three elements of urban form and their spatial configurations, with their established connections to earlier conceptualizations in urban morphology and urban ecology, represent a promising toolbox from which to design urban spatial systems that could carry and sustain not only social processes, but ecological processes in cities as well. Certainly this calls for further research, not least when it comes to developing these elements into performative systems, where the integration of space syntax and MatrixGreen seems especially promising.

BOX 4.3: ALBANO RESILIENT CAMPUS

There is a discernible movement in sustainable urban design from a first generation of research and practice, primarily addressing climate change, toward a second generation which broadens the field to also encompass biodiversity. The two have quite different implications for urban design and urban morphology. The first, stressing the integration of more advanced technological systems into the urban fabric, such as energy and waste disposal systems but also, more conspicuously, public and private transport systems, often leads to rather conventional design solutions albeit technologically enhanced. The second generation ask for a more direct involvement of urban form, asking the question: how are future urban designs going to harbor not only social and economic systems, which they have always done, but ecological ones as well? That is: how are we, as support for future practice in urban design, to develop knowledge that bridges the ancient dichotomy between human and ecological systems?

The project Albano Resilient Campus is to be regarded as a contribution to that second generation of practice. The project originated as an interdisciplinary research project between architects, urban morphologists, and systems ecologists supported by the public real estate company Akademiska hus, and was later adapted by the City of Stockholm in its formal plans for the area. It concerns a new university campus in Stockholm at the strategic site Albano located at the crossroads between three world-class institutions of higher education, with substantial potential to support a greater exchange between the faculties at the different universities through greater proximity and shared amenities. The site is also located at the border between the inner city with all its urban attractions and the Royal National City Park with unique natural and historic qualities. Thus, the aim has been to develop an urban campus that ties the established universities together and at the same time enhances the natural and ecological qualities in the area through the concept of social-ecological urban design.

Critical for this concept is to see how established design elements such as streets, building plots and buildings themselves can be made to support ecological systems and not only social ones. Just as social systems need connectivity between important nodes, so too do ecological systems, which is why strategic streets in this projects were conceived as "green arteries." Similarly, ecological systems, like social systems, need discrete spaces for the development of diverse uses (habitats), something conceived as "Active Ground" in the project. Finally, at certain locations eco-systems, like social systems, need to be intensified through particular constructions, such as water plants, allotment gardens, or green buildings. Such basic principles bring to the surface conflicting demands between social and ecological systems as well as generating deeper knowledge on the spatial performativity of ecosystems in an urban setting, which is why the project

Source: KIT Architects Stockholm.

has also given rise to collaborative research projects between researchers in Architecture at KTH and ecologists at Stockholm University aiming to develop a research field tentatively called Spatial Morphology of Urban Social-Ecological Systems.

4.6 DISCUSSION: DESIGN THEORY AS AN EPISTEMOLOGICAL FRAMEWORK

In this chapter, we have tried to identify the range of knowledge necessary for practice to successfully develop the urban fabric toward greater sustainability through urban design, by moving from the more speculative theories traditionally found in urban design (which we also believe typify most of the current literature in sustainable urban design) toward the many areas of relevance for urban design where more scientific, or analytical, theory is rapidly developing. We have also stressed the need for a more analytical theory development in spatial morphology, since we understand urban space as the prime medium in urban design.

Taken together, this firmly positions our endeavor in line with recent developments in design theory, which argues that it is exactly through a deeper epistemological understanding that the design process can be demystified (Lawson, 2005). The design process can be fruitfully described as a process wherein different knowledge forms are applied (Marcus, 2010b). First, in any design process we need support in finding potential solutions to a design problem, and this is exactly what speculative theory of the kind discussed earlier offers. A typical case of speculative theory is *theory in art* (e.g. *The Futurist Manifesto*); that is, theory that offers new ways to look at the world thus generating new possibilities – in the futurist case, the world as movement, generating new expressive possibilities in art (Hillier, 1996). Typical of the successful designer is her or his ability to assimilate a wide range of such ways to look at the world, a *repertoire* (Schön, 1983) from which she or he can search for possible solutions given a particular design problem. Second, possible solutions obviously need testing in terms of their appropriateness for the task at hand, and it is exactly at this moment in the design process that there is a need for *analytical* knowledge that can tell the designer, given the demands of the particular case, whether the proposed solution will work or not. A typical speculative theory in architecture – such as Le Corbusier's *la ville contemporaine*, which (to simplify greatly) promised housing in large, tranquil green landscapes – clearly needed testing. Such a need is evident in our wide experience of the unexpected outcome of this concept – for example, when it came to both desolation and traffic congestion. This does not diminish Le Corbusier's remarkable achievement in generating a vast new field of possibility in urban planning and design, but it does underscore the need for analytical theory that can test claims made by speculative theory. However, analytical knowledge can only answer a limited range of questions raised by a proposed design – in particular, questions that can be quantified. This is why any design proposal also needs to be addressed by *discursive* knowledge – that is, knowledge that can contextualize and evaluate a proposal in its wider social and cultural setting, knowledge we typically find in the humanities or the social sciences. While we have extensively discussed the two first knowledge forms here, speculative and analytical knowledge, we have not touched upon the last, discursive knowledge, but have instead relied on contributions from our colleagues in adjoining chapters. The main argument in this chapter has been that, given current global challenges, in order to develop more sustainable urban fabrics in our cities we need to expand the knowledge base in the professional practices in the field in order to also incorporate analytical theory to a much greater extent than today, and we must do so without losing track of the traditional strength of the field in speculative theory.

REFERENCES

Ahrné, K. (2008) "Local management and landscape effects on diversity of bees, wasps and birds in urban green areas," Doctoral thesis No. 2008:41, Faculty of Natural Resources and Agriculture Science, Swedish University of Agricultural Science (SLU).

Alberti, M. (2005) "The effects of urban patterns on ecosystem function," *International Regional Science Review*, 28(2): 168–192.

Alberti, M. (2008) *Advances in Urban Ecology: Integrating Humans and Ecological Processes in Urban Ecosystems*, New York: Springer.

Andersson, E. and Bodin, Ö. (2009) "Practical tool for landscape planning? An empirical investigation of network based models of habitat fragmentation," *Ecography*, 32: 123–132.

Andersson, E., Barthel, S., and Ahrné, K. (2007) "Measuring social-ecological dynamics behind the generation of ecosystem services," *Ecological Applications*, 17: 1267–1278.

Balfors, B. and Mörtberg, U. (2006) "Landscape Ecological Assessment: A tool for prediction and assessment of impacts on biodiversity," paper given at the conference "Ecological Impact Assessments: Science and Best Practice," Bath Spa University, 11–12 July, British Ecological Society.

Balfors, B., Mörtberg, U., Gontier, M., and Brokking, P. (2005) "Impacts of region-wide urban development on biodiversity in strategic environmental assessment," *Journal of Environmental Assessment Policy and Management*, 7: 229–246.

Balfors, B., Mörtberg, U., and Geneletti, D. (2010) "Landscape ecology for SEA: Lessons learned," paper given at the 30th Annual Conference of the IAIA, "The Role of Impact Assessment in Transitioning to the Green Economy," 6–11 April, Geneva.

Barthel, S., Folke, C., and Colding, J. (2010a) "Social-ecological memory in urban gardens: Retaining the capacity for management of ecosystem services," *Global Environmental Change*, 20: 255–265.

Barthel, S., Colding, J., Erixon, H., Ernstson, H., Grahn, S., Kärsten, C., Marcus, L., and Torsvall, J. (2013). *Principles of Social-Ecological Urban Design: The Case of Alban University Campus*, TRITA-ARK-2013:3, KTH, Stockholm, Sweden.

Batty, M. (2005) *Cities and Complexity*, Cambridge, MA: MIT Press.

Beatley, T., Newman, P., and Boyer, H. (2009) *Resilient Cities: Responding to Peak Oil and Climate Change*, Washington, DC: Island Press.

Bernow, R. and Ståhle, A. (2011) "Värdering av stadskvaliteter: PM Sammanfattning av metod och resultat," Stockholms stad, Stockholms Läns Landsting, Haninge kommun, Lidingö stad, Nacka kommun, Stockholm.

Björkman, L.-L. (2012) *Fritidsodlingens omfattning i Sverige*, Rapport 2012:8, Alnarp: Swedish University of Agricultural Sciences, Fakulteten för landskapsplanering, trädgårds- och jordbruksvetenskap.

Bodin, Ö. and Zetterberg, A. (2010) *MatrixGreen v 1.6.4 User's Manual: Landscape Ecological Network Analysis Tool* (www.matrixgreen.org), Stockholm: Stockholm Resilience Centre and KTH Royal Institute of Technology.

Bolund, P. and Hunhammar, S. (1999) "Ecosystem services in urban areas," *Ecological Economics*, 29: 293–301.

Borgström, S. T., Elmqvist, T., Angelstam, P., and Alfsen-Norodom, C. (2006) "Scale mismatches in management of urban landscapes," *Ecology and Society*, 11(2): art. 16.

Brito, L. and Stafford Smith, M. (2012) "State of the planet declaration," in *Planet under Pressure: New Knowledge towards Solutions*, International Council for Science.

Calthorpe, P. (2010) *Urbanism in the Age of Climate Change*, Washington, DC: Island Press.

Choay, F. (1997) *The Rule and the Model*, Cambridge, MA: MIT Press.

Choi, E. (2012) "Walkability as an urban design problem: Understanding the activity of walking in the urban environment," Stockholm: KTH.

Colding, J. (2007) "'Ecological land-use complementation' for building resilience in urban ecosystems," *Landscape and Urban Planning*, 81: 46–55.

Colding, J., Lundberg, J., and Folke, C. (2006) "Incorporating green-area user groups in urban ecosystem management," *Ambio*, 35: 237–244.

Congress for the New Urbanism (1999) *Charter of the New Urbanism*, New York: McGraw-Hill Professional.

Corner, J. (1999) *Recovering Landscape: Essays in Contemporary Landscape Theory*, Princeton, NJ: Princeton Architectural Press.

Costanza, R., d'Arge, R., de Groot, R., Farber, S., Grasso, M., Hannon, B., Limburg, K., Naeem, S., O'Neill, R. V., Paruelo, J., Raskin, R. G., Sutton, P., and van den Belt, M. (1997) "The value of the world's ecosystem services and natural capital," *Nature*, 387: 253–260.

Desyllas, J. (2000) "The relationship between urban street configuration and office rent patterns in Berlin," PhD thesis, University College London.

Dittmar, H. and Ohland, G. (2003) *The New Transit Town: Best Practices in Transit-Oriented Development*, Washington, DC: Island Press.

Droege, P. (2007) *The Renewable City: A Comprehensive Guide to an Urban Revolution*, New York: John Wiley.

Duany, A., Speck, J., and Lydon, M. (2009) *The Smart Growth Manual*, New York: McGraw-Hill Professional.

Elmqvist, T., Folke, C., Nyström , M., Peterson, G., Bengtsson, J., Walker, B., and Norberg, J. (2003) "Response diversity, ecosystem change, and resilience," *Frontiers in Ecology and the Environment*, 1: 488–494.

Ernstson, H., Barthel, S., Andersson, E., and Borgström, S. T. (2010) "Scale-crossing brokers and network governance of urban ecosystem services: The case of Stockholm," *Ecology and Society*, 15(4): art. 28.

Farr, D. (2007) *Sustainable Urbanism: Urban Design with Nature*, New York: John Wiley.

Folke, C. (2006) "Resilience: The emergence of a perspective for social-ecological systems analysis," *Global Environmental Change*, 16(3): 253–267.

Gill, S. E., Handley, J. F., Ennos, A. R., and Pauleit, S. (2007) "Adapting cities for climate change," *Built Environment*, 33(1): 115–133.

Glasson, J., Therivel, R., and Chadwick, A. (1999) *Introduction to Environmental Impact Assessment: Principles and Procedures, Process, Practice and Prospects*, London: UCL Press.

Gontier, M., Balfors, B., and Mörtberg, U. (2006) "Biodiversity in environmental assessment: Current practice and tools for prediction," *Environmental Impact Assessment Review*, 26: 268–286.

Gontier, M., Mörtberg, U., and Balfors, B. (2010) "Comparing GIS-based habitat models for applications in EIA and SEA," *Environmental Impact Assessment Review*, 30(1): 8–18.

Greene, M., Reyes, J., and Castro, A. (2012) *Proceedings of the Seventh International Space Syntax Symposium*, Pontefica Universidad Católica de Chile.

Haas, T. (2008) *The New Urbanism and Beyond: Designing Cities for the Future*, New York: Rizzoli.

Haas, T. (2012) *Sustainable Urbanism and Beyond: Rethinking Cities for the Future*, New York: Rizzoli.

Hildén, M., Valve, H., Jónsdóttir, S., Balfors, B., Faith-Ell, C., Moen, B., Peuhkuri, T., Schmidtbauer, J., Swensen, I., and Tesli, A. (1998) *EIA and Its Application for Policies, Plans and Programmes in Sweden, Finland, Iceland and Norway*, TemaNord 1998:567, Copenhagen: Nordic Council of Ministers.

Hillier, B. (1996) *Space Is the Machine*, Cambridge: Cambridge University Press.

Hillier, B. and Hanson, J. (1984) *The Social Logic of Space*, Cambridge: Cambridge University Press.

Hillier, B. and Iida, S. (2005) "Network effects and psychological effects: A theory of urban movement," in A. Cohn and D. Mark (eds.) *Spatial Information Theory*, Berlin: Springer Verlag, pp. 473–490.

Hillier, B. and Xu, J. (2004) "Can streets be made safe?" *Urban Design International*, 9: 31–45.

Hillier, B., Penn, A., Hanson, J., Grajewski, T., and Xu, J. (1993) "Natural movement: or, configuration and attraction in urban pedestrian movement," *Environment and Planning B: Planning and Design*, 20: 29–66.

Holling, C. S. (1973) "Resilience and stability of ecological systems," *Annual Review of Ecology and Systematics*, 4: 1–23.

Kennedy, C., Cuddihy, J., and Engel-Yan, J. (2007) "The changing metabolism of cities," *Journal of Industrial Ecology*, 11: 43–59.

Koch, D. and Steen, J. (2012) "Decomposing programmes: Re-coding hospital work with spatially syntactic information," in M. Greene, J. Reyes, and A. Castro (eds.) *Proceedings of the Eighth International Space Syntax Symposium*, Santiago de Chile: PUC.

Koch, D., Marcus, L., and Steen, J. (2009) *Proceedings of the Seventh International Space Syntax Symposium*, Trita-Ark-Research Publication 2009:1, KTH.

Lawson, B. (2005) *How Designers Think: The Design Process Demystified*, Oxford: Architectural Press.

Legeby, A. (2010) "Urban segregation and urban form: From residential segregation to segregation in public space," Licentiate thesis, TRITA-ARK 2010:1, KTH, Stockholm.

Lindström, P. and Lundström, M. J. (2008a) "Provide sustainable environmental solutions in urban development," in P. Lindström and M. J. Lundström (eds.) *Sustainability by Sweden: Perspectives on Urban Governance*, Karlskrona, Sweden: Boverket, pp. 26–29.

Lindström, P. and Lundström, M. J. (2008b) "SymbioCity: The holistic approach to sustainable urban development," in P. Lindström and M. J. Lundström (eds.) *Sustainability by Sweden: Perspectives on Urban Governance*, Karlskrona, Sweden: Boverket, pp. 4–10.

Marcus, L. (2000) "Architectural knowledge and urban form," PhD thesis, TRITA-ARK 2000:2, KTH, Stockholm.

Marcus, L. (2010a) "Spatial capital: A proposal for an extension of space syntax into a more general urban morphology," *Journal of Space Syntax*, 1: 30–40.

Marcus, L. (2010b) "The architecture of knowledge for educations in Urban Planning and Design," *Journal of Space Syntax*, 1: 214–229.

Marcus, L. and Colding, J. (2011) "Towards a spatial morphology of urban social-ecological systems," paper presented at "Urban Morphology and the Post-carbon City," ISUF 2011 (the 19th International Seminar on Urban Form), Montreal, 26–29 August.

Millennium Ecosystem Assessment (MEA) (2005) *Ecosystems and Human Well-Being: Synthesis*, Washington, DC: Island Press.

Mörtberg, U., Balfors, B., and Knol, W. C. (2006) "Landscape ecological assessment: A tool for integrating biodiversity issues in strategic environmental assessment and planning," *Journal of Environmental Management*, 82(4): 457–470.

Mörtberg, U., Zetterberg, A., and Balfors, B. (2012) "Urban landscapes in transition, lessons from integrating biodiversity and habitat modelling in planning," *Journal of Environmental Assessment Policy and Management*, 14: 1250002-1–1250002-30.

Mostafavi, M., Doherty, G., and Harvard University Graduate School of Design (2010) *Ecological Urbanism*, Zürich: Lars Müller.

Netzell, O. (2010) "Essays on lease and property valuation," PhD thesis, TRITA-FOB 2010:4, KTH, Stockholm.

Palazzo, D. and Steiner, D. (2011) *Urban Ecological Design*, Washington, DC: Island Press.

Ranhagen, U. and Groth, K. (2012) *The SymbioCity Approach*, Stockholm: SKL International.

Rutherford, J. (2008) "Unbundling Stockholm: The networks, planning and social welfare nexus beyond the unitary city," *Geoforum*, 39: 1871–1883.

Schön, D. (1983) *The Reflective Practitioner*, New York: Basic Books.

Scoppa, M. (2012) "Towards a theory of distributed attraction: The effects of street network configuration upon the distribution of retain in the City of Buenos Aires," dissertation, The School of Architecture, Georgia Institute of Technology, Atlanta, Georgia, USA.

Shan, Y., Jingping, C., Liping, C., Zhemin, S., Xiaodong, Z., Dan, W., and Wenhua W. (2007) "Effects of vegetation status in urban green spaces on particle removal in a street canyon atmosphere," *Acta Ecologica Sinica*, 11: 4590–4595.

Simon, H. (1969) *Sciences of the Artificial*, Cambridge, MA: MIT Press.

Ståhle, A. (2005) "Mer park i tätare stad: Teoretiska och empiriska undersökningar av stadsplaneringens mått på friytetillgång," Licentiate thesis, TRITA-ARK 2005:2, KTH, Stockholm.

Ståhle, A. (2008) "Compact sprawl: Exploring public open space and contradictions in urban density," PhD thesis, TRITA-ARK 2008:6, KTH, Stockholm.

Ståhle, A., Marcus, L., and Karlström, A. (2005) "Place syntax: A space syntax approach to accessibility," in *Proceedings of the Fifth International Space Syntax Symposium*, Delft University of Technology, pp. 131–139.

Therivel, R. (2010) *Strategic Environmental Assessment in Action*, 2nd ed., London: Earthscan.

Vaughan, L. (2007) "The spatial syntax of urban segregation," *Progress in Planning*, 67: 205–294.

Waldheim, C. (2006) *The Landscape Urbanism Reader*, Princeton, NJ: Princeton Architectural Press.

Wilkinson, C. (2012) "Social-ecological resilience and planning: An interdisciplinary exploration," doctoral thesis in Natural Resources Management, Department of System Ecology, Stockholm University.

Zetterberg, A. (2011). "Connecting the dots: Network analysis, landscape ecology, and practical application," PhD thesis, Trita-LWR, ISSN 1650-8602; 1062, KTH, Stockholm.

Zetterberg, A., Mörtberg, U., and Balfors, B. (2010) "Making graph theory operational for landscape ecological assessments, planning and design," *Landscape and Urban Planning*, 95(4): 181–191.

CHAPTER 5

SUSTAINABLE URBAN FLOWS AND NETWORKS

THEORETICAL AND PRACTICAL ASPECTS OF INFRASTRUCTURE DEVELOPMENT AND PLANNING

Folke Snickars,
Lars-Göran Mattsson,
and Bo Olofsson

5.1 INTRODUCTION

CRITICAL URBAN INFRASTRUCTURE provides the preconditions for urban life. This chapter discusses the role of urban sustainability in relation to infrastructure systems such as the road network, public transportation, telecommunications networks, and water and waste management. In particular, the planning and management systems for infrastructure provision in the Stockholm region are analyzed in relation to sustainability. Similarities and differences among different types of urban infrastructure networks are compared, as well as how they interact, co-evolve, and sometimes counteract each other. The chapter also investigates the environmental and social effects of the strategic choices made with regard to infrastructure systems development.

We first discuss the concept of *infrastructure* and its interpretations in urban and regional infrastructure networks. We then develop a conceptual model of strategic planning as a process of disentangling time scales and handling uncertainty. These theoretical considerations are then brought to bear on the issue of sustainable infrastructure provision in the Stockholm region. An important distinction is made between infrastructure provision and market solutions to the creation of services using infrastructure systems. We also discuss whether or not a region like Stockholm can continue growing without extending its current infrastructure systems, and question who will plan and finance such systems under a growth scenario.

The Stockholm region can be seen as an economic and social system built on both natural endowments and man-made infrastructure systems. Originally, Lake Mälaren was a brackish inner archipelago of the Baltic Sea. As a consequence of the phenomenon of land upheaval (the slow resurfacing of land that was pressed down beneath sea level by glacial expansion during the last Ice Age), the natural flow interchange between Lake Mälaren and the Baltic Sea was blocked, and a freshening process began in Lake Mälaren. Ever since, the fresh water from the lake discharges, meets, and mixes with the brackish water of the Baltic Sea in the very center of the city, and owing to the resulting difference in water levels between the lake and the sea, from early medieval times ships had to be towed past the rapids in front of what is now the Old Town.

This passage, the present-day sluice, *Slussen*, also had to be protected and defended to maintain control of the lands surrounding Lake Mälaren. Stockholm's first castle was built on an outcrop on the southern side of the rapids. The land has continued to rise 3–4 mm/year since Stockholm was founded, during the medieval period, but owing to natural erosion of the sand and gravel deposit forming the Old Town, the water level of Lake Mälaren is still less than half a meter higher than that of the Baltic Sea. The freshening of Lake Mälaren made it possible to use it as fresh water supply, which was an important precondition for the establishment of the city. The water supply of the entire population of the Stockholm region – over 2 million people – still comes from the surface water of the lake. This is a great natural asset of Stockholm but also makes the city vulnerable to both human and natural threats. For example, if climate change raises seawater levels (even only slightly), the difference in water levels in Lake Mälaren and the Baltic Sea would be reversed, leaving Stockholm without a reliable fresh water supply.

Stockholm is an example of a metropolitan region where natural endowments are important determinants of both (radial) transportation systems (road and rail) and most utility systems. Geography has contributed to creating an unusually strong concentration of activity in a spatially relatively limited inner city. For example, Stockholm's natural endowment of fresh water, with the sluice in front of the Old Town, has shaped the development of the city's urban structure and built environment, including its networks for transportation, water, sewage, and energy. The water system acts both as a physical barrier and as an urban amenity (and, historically, a transport amenity) for the city's inhabitants.

Stockholm is, at the same time, a region that is greatly dependent on its transportation system. It has a lower density than one might expect for

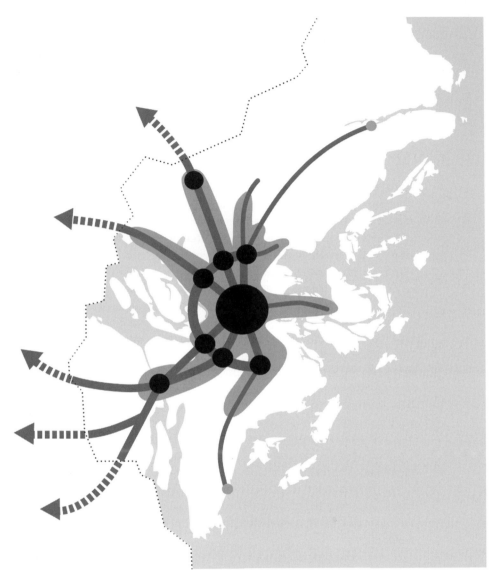

Figure 5.1 Stylized picture of the Stockholm region and its development directions
Stockholm County Council, 2010

a region of its size, but development is concentrated along transport corridors. As is illustrated in Figure 5.1, the region has a distinct finger structure,
with suburban centers located at the knuckles of the fingers. One reason for
this organized dispersal is the subway and commuter train systems, which
provide the bones of the fingers and extend a considerable distance from
the central business district. The decentralized building stock is also a reflection of decentralized decision making with respect to building, planning, and

public service provision. Around the City of Stockholm, there are 24 other municipalities in the region with independent powers to tax citizens for public services ranging from schools to old age care, energy provision, and waste disposal.

The geography, natural resources, and institutional structure of the Stockholm region thus make it more dependent on the efficiency of its networks than a region built on a continuous plain, with a single joint political decision-making body. The Stockholm region is a complex urban system in which political decision makers must simultaneously balance strategies of cooperation and competition to promote growth and sustainability. The sustained existence of institutions for regional planning in the Stockholm region is a reflection of this complexity. The Stockholm County Council has had political responsibility for planning at the metropolitan level since 1971, and this includes responsibility for planning and operating public transport and health care. It shares responsibility for a range of other types of infrastructure provision with political bodies at both the national and the municipal levels. However, most regional coordination of policy is conducted on a voluntary basis by municipalities as a consequence of the planning monopoly accorded to Swedish municipalities by the Swedish constitution. The planning monopoly has strong democratic merits but also leads to challenges such as undesirable scale effects at the regional level due to uncoordinated municipal decision making, and what could be seen as suboptimal regional solutions with regard to land-use planning and the related potential for the sustainable expansion of infrastructure systems.

5.2 WHAT IS INFRASTRUCTURE?

In the 1980s and 1990s, a number of Swedish urban researchers argued that any metropolitan region (e.g. the Stockholm region) could be seen as a form of physical, economic, and social infrastructure (Andersson *et al.*, 1984; Andersson and Strömqvist, 1987; Johansson and Snickars, 1992). This conceptualization defines infrastructure as a variant of material or non-material capital (see, for example, Johansson and Snickars, 1992). The most important property of infrastructure is versatility in use, from different time and activity perspectives. In this sense, this concept of infrastructure is strongly related to the notion of sustainable development: both concepts emphasize a long-term perspective to guide development strategies.

One way to define infrastructure is in terms of technical properties. Table 5.1 summarizes some of the most important types of urban infrastructure from such a perspective, including networks and also building stocks. If infrastructure is seen as the arena for urban activities, one can say that the urban area is itself an infrastructure that is, in turn, also made up of

Table 5.1 Urban infrastructure as technical and organizational artifacts

Utility supply networks	Information and communication networks	Transport networks	Building stocks
Water distribution	Telephone	Roads and streets	Housing areas
Electricity and gas	Post	Freeways	Office buildings
District heat	Cable television	Rail, subway, and	Commercial service
Waste handling	Traffic control	tram tracks	buildings
Sewerage	Broadband internet	Air corridors	Education facilities
		Harbors, fairways	Hospitals
		and canals	Culture facilities
		Pipes	Urban environments

infrastructures. A number of researchers have tried to assess the total value of the infrastructure capital of an urban region. An example of this is given in Hudson *et al.* (1997), which addresses New York, indicating the enormous amount of capital stock involved. The capital stocks are measured both in physical terms and in value terms, and a detailed subdivision into networks, utility installations, and buildings is provided.

It is instructive to complement artifact-oriented definitions with temporal, spatial, and functional ones (see Table 5.2). Focusing on the *temporal dimension*, we can define infrastructure as that form of capital which has a spatial fixity across time. Capacities change slowly over time, and the process of adaption between economic and social processes and infrastructure develops with time lags. From this perspective, one can see infrastructure as a stable or slowly changing platform, or arena, upon which faster processes operate. This gives rise to nested dynamic processes that interact with each other across time. Planning often involves complex

Table 5.2 Properties distinguishing infrastructure capital from other forms of capital

Scale and scope	Properties defining infrastructure
Temporal	Durable capital with fixed location
	Slowly changing capacity and spatial pattern
	Slow speed compared to social and economic activity processes
	Stable platform on which fast processes operate
Spatial	Services offered to a spatial configuration
	Services offered to a collective of users
	Utility normally decreases with distance
Functional	Basic feature is that of creating accessibility
	First fundamental is versatility in use
	Second fundamental is generality across time
	Third fundamental is system and network property
	Fourth fundamental is indivisibility

interactions among the fast and slow processes in urban areas; thus, when new additions to the road or rail networks are planned, equilibrium is assumed in the rapidly changing traffic flows. Inversely, when operations are planned in the public and private traffic systems, inert capital stocks are fixed.

From a *spatial perspective*, infrastructure services are offered to a geographically distributed configuration of users. The utility to users normally decreases with distance. The first spatial property of infrastructure is often termed its *network property*. In other words, the utility of infrastructure is related to its ability to facilitate interaction among different places. A spatial perspective also draws attention to the relationship between private and public goods with respect to infrastructure. This is conceptually different from the idea of a spatially fixed public good, although there are rather strong connections between the concepts; see Johansson and Snickars (1992). A good can be *public* on either the supply side or the demand side, or both. If it is a *local* public good, it will confer utility only locally or within an administratively defined territorial unit. A subway system is an example of a *regional* public good, even if it can be argued to have rivalrous properties due to decreased user utility in cases of congestion. The Stockholm National City Park and the green corridors between the development fingers of the Stockholm region are also regional public goods (see Figure 5.1). Sports facilities are often local public goods, whereas the distribution network for electricity is a national and to some extent international public good. Since network services are primarily provided to the Stockholm region's inhabitants, one can argue that they should rather be termed regional public goods – or *club goods*.

Finally, we can think in terms of the *functional properties* of infrastructure – that is, that infrastructure creates the potential for interaction, described by geographers such as Hägerstrand (1973) as *accessibility*. The first fundamental functional property of infrastructure is versatility in use. A road can, for instance, be used to serve a variety of different activities at a single moment in time. The second fundamental functional property of infrastructure is generality across time. A road stays fixed for a long period, making it possible for current users to take both short- and long-term actions with the confidence that the opportunity to use the road will remain.

The third fundamental functional property is the *system-and-network* property of infrastructure. A single new road connection linking two unconnected parts of the city creates a synergy; it opens up new opportunities. The fourth fundamental functional property of infrastructure is indivisibility. In other words, it is often impossible to supply infrastructure that is perfectly adapted to demand. Rather, many infrastructure investments are "lumpy" – they are by necessity provided in increments. In the urban context, this may, for instance, mean that investments in infrastructure will affect (and be financed by) many people who do not see the investment as being in their individual best interest. But taken as a whole, the collective group of potential users will all be positively affected by the cost reduction made possible by the provision of new infrastructure (a new subway link is a good example). However, despite this general tendency certain forms of infrastructure investment do have uneven distribution and welfare effects (see Graham and Marvin, 2001).

Fast and slow processes of urban systems

It may be instructive to distinguish between fast and slow processes at the micro and macro levels in order to understand the complexity of urban infrastructure dynamics (also see the discussion in Bettencourt *et al.*, 2007, on the paces of urban life). Figure 5.2 gives some examples of slow and fast processes at the individual or micro scales, and at larger or systemic scales. In practice, of course, the more interesting question is how slow and fast processes interact with micro and macro scales, and how such interactions give rise to systemic patterns and dynamics. For example, processes could be slow at the micro level and fast at the macro level; the slow individual process of moving from one dwelling to another at the micro level can give rise to a fast process of housing market dynamics at the macro level, as many people make similar decisions. By contrast, "fast" daily individual decisions to commute to work at a specific time, using a specific transport mode, lead to a predictable (slow-changing) congestion pattern at the macro level.

Economies of scale and scope in a metropolitan region will, according to established arguments, generate the economic surplus necessary to extend infrastructure systems so that net benefits remain positive even during periods of rapid urban growth. In other words, even if infrastructure provision is lumpy, individual assessments of benefit and efficiency change quickly and populations grow and/or become more diverse. Hence, large regions can provide infrastructure in increments that benefit the population as a whole. If we accept this public goods argument, then metropolitan regions should be able to provide infrastructure at lower per capita costs with ensuing positive consequences for sustainability. Thus, because of this markedly higher efficiency of use, sustainability at the national level would be promoted by a more rapid urbanization process.

Infrastructure *development* is often determined according to a planning principle or, to put it differently, a principle of collective choice (see Törnqvist, 2011). Infrastructure *use*, by contrast, is normally guided by market principles. One argument often brought to bear in considering planned public provision vs. market provision is that actions in one period may have irreversible effects in a future period. The urban system is a complex and dynamic system. Just as small changes in global temperature can have catastrophic systemic effects beyond a "tipping point," so too can city systems have tipping points that lead to irreversible changes in conditions for future development. An example of such a tipping point is the process of the

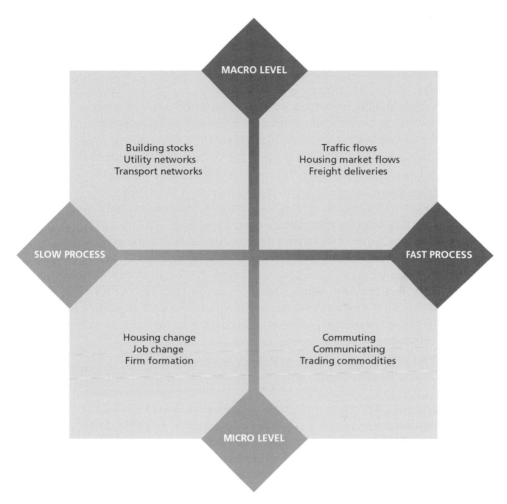

Figure 5.2 Fast and slow urban processes at micro and macro scales

degradation of housing estates. A slow process wherein individual house-holds change their choice of dwelling, say moving out of a particular housing estate one by one, may lead to a rapid process of deterioration – first in the status of the area, and then in its physical appearance. Once this new state has emerged, it is very difficult to reverse (see also Olsson, 2002).

Infrastructure planning is at the core of regional planning, owing to its durability and systemic complexity, and the long-term nature, uncertainty, and risk associated with its effects. The basic planning problem is to choose a policy package (say, a subway system) that maximizes a welfare function (such as long-term sustainability) under the best possible forecast of external events (such as national economic growth). Planners can base infrastructure decisions on scientific knowledge of the system's function, but also need to consider the ways in which cooperation with market actors can affect scenarios regarding things like network vulnerability and public transport ridership. All of these decisions reflect different types of uncertainties about the implications of planning processes and public and private investments in infrastructure.

Planners and decision makers face many different kinds of uncertainty in approaching infrastructure investments. The nature of uncertainty depends on the apprehended properties of the specific systems in focus. The problem is that, most of the time, planners cannot be completely certain about the nature of uncertainties, but can only make educated guesses or develop framing assumptions. For instance, they may sometimes reason that the uncertainties relating to a specific system are *dynamic* – that is, that the level of inputs and outputs, and states, change over time – but that the mechanisms of the system stay relatively stable. Such dynamic uncertainty dissolves over time, allowing planners to make incremental changes based on current information: an adaptive decision strategy. A typical situation is a government budget decision, which, though based in part on long-term forecasts, appropriates funds on the basis of current conditions. Or, systems can have inputs that are relatively stable or static, but the system framework itself may be dynamic – that is, the mechanisms can evolve or change. The result is what can be called *quasi-static uncertainty*, and in such situations it pays to constantly buy information to reduce risk and stay vigilant to emerging changes. For example, public authorities may commission university research to improve the informational basis for decision making.

If we regard uncertainty as *static* (i.e. ever-present), a flexible decision strategy may be best, to avoid regret when uncertain external factors become known. This is the main idea behind the so-called *precautionary principle*. A typical situation is climate change at the global level. Since static uncertainty will not be resolved over time, the appropriate action to take is to reduce vulnerability to uncertain future external events. This strategy might be applied in designing district heating networks that will work well both with current feedstocks and also with alternative energy sources, like renewables, that may be more prevalent in the future. Another example would be to restrict new urban construction to areas less likely to be flooded in the event of climate change. These notions have been placed in the context of vulnerability by Berdica (2002), Jenelius *et al.* (2006), and Berdica and Mattsson (2007). Most examples in this literature study transport systems but provide insights relevant to most other urban infrastructure systems.

Finally, *competitive uncertainty* is present when decision makers are strategically interdependent. At the regional level, different actors often act strategically against one another. For instance, Swedish municipalities compete for households, and therefore make strategic decisions in the area of housing construction to improve their competitive position in relation to neighboring municipalities.

In the anticipation of broad, or genuine, uncertainty of multiple types, using scenarios is a way of narrowing down the decision options. There are two interpretations of scenarios. One definition associates the term "scenario" with external factors such as those described as "external events" in Table 5.3. Scenarios may also denote possible states of the system as a whole. In this mode of thinking, all of the elements together comprise one scenario.

5.3 CHALLENGES IN STOCKHOLM'S INFRASTRUCTURE SYSTEMS

Assuring a long-term sustainable urban fabric demands robust yet flexible infrastructure systems for the movement of people and goods, the supply of water, waste handling, energy systems, and telecommunications. Table 5.3 illustrates some components of planning action in the provision of infra-structure for sustainability in the Stockholm region. We imagine a situation in which the supply of infrastructure is mainly a public concern and the supply of services is a concern of a combination of public and private actors. A set of conceivable planning instruments and investment objects available to actors in the region are listed in the first column. A number of responses from

Table 5.3 Some components of infrastructure-led scenarios for the Stockholm region

Infrastructure system	Planning instruments	Market outcomes	External event
Roads	Network investments and management Pricing instruments	Car purchases and use Fuel provision	Random-access behavior Job specialization and work practices
Tracks	Network investments and multimodal exchanges Ticketing frameworks	Competitive service providers Intelligent guidance systems	Ridership practices Tax subsidy systems
Energy	District heat systems Smart electricity grids Reliability requirements	Service duopoly systems Renewable fueling	Energy conservation practices Price paths and climate change
Water and waste	Closed-loop systems Local subsystems	Municipal roles Life-cycle considerations	Demand development Recycling practices
Telecommunications	Cloud computing and the internet Wiring the building stock	Productivity of service providers Business climate paths	Digital lifestyle development E-democracy developments

the market actors in the urban infrastructure field are given in the second column. Examples of external events to the infrastructure market actors that may have an impact on the outcome of the combined actions of planning agencies and market actors are noted in the third column. These are predominantly actions of the users of the infrastructure services among the Stockholm region's firms and households. In the third column, we have also listed possible policy packages at the national level affecting the choices made by regional actors.

The fundamental question is how actions taken in the infrastructure systems, either by planning actors, by market actors on the supply side, or by the households and firms using the services, will affect the sustainability of the future Stockholm region. Should Stockholm's infrastructure systems be extended to meet changing and growing demand, or can existing systems be better managed? Can sustainability be promoted, even within the existing infrastructure systems, using careful renewal strategies for existing nodes and networks? The following subsections give some examples of critical infrastructure system decisions currently under discussion in the Stockholm region.

Water and sewage systems

Several programs to ensure that Stockholm retains a sustainable supply of water have been implemented or are ongoing. Networks have been established around the City of Stockholm: 11 municipalities in the southern part of the region as well as the 14 municipalities comprising the Norrvatten Association north of Stockholm have been connected in order to increase supply security. Another prime example of a project aimed at improving sustainability is the water protection area established in the eastern part of Lake Mälaren in 2008, which also comprises parts of the upstream catchment area. Long-established land-use restrictions around Lake Bornsjön, the main water reservoir backup for the City of Stockholm, are so effective that the lake's water is directly potable without cleaning.

Projects and programs for high-grade cleaning of wastewater have been successively introduced, not only in the Stockholm region but in all urban settlements around Lake Mälaren. These have been so successful that the water quality is good enough for swimming even in the central areas of the City of Stockholm. Furthermore, the treated wastewater from northwest Stockholm is directed through a tunnel system under the City of Stockholm to the Baltic Sea to prevent it from entering the city's water system. Heat is extracted from the wastewater stream before the water is released into the sea, a process that provides heat energy and also contributes to better mixing of water bodies during the winter months.

Massive investments have been made in the management of black water (wastewater containing excrement) in the Stockholm region, through both local and regional projects and systems. The downtown area of the City of Stockholm has been provided with an integrated system for wastewater and black water. A tunnel system has been constructed that functions as a reservoir to avoid flooding of the combined system during heavy rainfall. Such rains are most likely to become more common in both the near and the distant future as a result of climate change, so there will be a need to introduce additional risk management systems in the future. Local solutions and development projects under discussion include different forms of wetlands, sedimentation basins, and oil-separating facilities.

Some of the most important challenges in Stockholm's continued infrastructure development relate to climate change. Climate change is very likely to change the character of water-level fluctuations in the downtown area, for instance in the Old Town. These fluctuations will also cause flooding in the wastewater systems, which are not dimensioned for such heavy rainfall. Among other things, this might require speeding up the introduction of refurbishments to the central sluice (Slussen), which has remained untouched for a considerable time. Otherwise, there will be an increased risk of flooding in the underground sections of the subway system. In the short term, because of a changed precipitation pattern the critical issue is to increase the discharge capacity of water from Lake Mälaren to the Baltic Sea. The reconstruction of the sluice at Slussen is therefore an urgent and necessary adaptation to climate change. However, in the long run, water will have to be pumped out of Lake Mälaren to hold back salt water from flowing backwards through the sluice area, a difficult task indeed for the downtown area as a whole. The only viable long-term solution is a wall and a pumping system placed well outside of the city center to block salt water.

Energy systems

In the area of sustainable energy flows, the Arlanda Airport aquifer system is a novel best-practice example of a more localized systemic solution, being the largest such underground storage system in the world. Figure 5.3 illustrates a novel way of designing closed-loop systems for energy management at one of the most important transport facilities in the Stockholm region: Arlanda International Airport. The energy management system has been in service since the summer of 2009. Arlanda consumes as much energy as a city of 25,000 people. An aquifer located in a gravel deposit nearby the airport helps make the heating and cooling of the airport area both inexpensive and environmentally friendly. The aquifer can be described as a huge groundwater reservoir. All heating and cooling of airport buildings (including the terminals) will, once the system is fully complete, come from the aquifer. Cold water is pumped out of the aquifer in the summer and used in the airport's district cooling network. Warmed-up water then flows back and is pumped underground and stored until winter, when it is needed to

Figure 5.3 Closed-loop system for ground heat extraction and heating/cooling management at Arlanda Airport

melt the snow in aircraft parking stands and pre-warm the ventilation air in the terminals and outbuildings.

There are many ongoing and planned similar investments to increase the sustainability of energy supply systems on a larger scale. For example, in central Stockholm groundwater is circulated for both heating and cooling through the gravel ridge that runs through the city; this system serves central neighborhoods such as the Old Town and the main north–south street, Sveavägen. It is very common today to drill into Stockholm's granite hills to install rock-based heat pumps, even in the downtown area of Stockholm.

Indeed, the structural conflict between district heating systems and individual building energy solutions provides an interesting example of a dilemma between public and private interests in sustainable urban infra-structure systems. District heating can efficiently deliver heating (and in some cases cooling) to urban areas, provided that there are many user con-nections with relatively short distances between them. However, individual property owners may find it more attractive to invest in individual building solutions that offer low energy costs, such as heat exchange pumps. The question is also how we should look upon the district heating system of the City of Stockholm from a sustainability perspective, since that system relies partly upon fossil fuels. Research-based proposals have been put forward to increase efforts in the field of geothermal energy for district heating, but it is difficult to promote such ideas because the regional energy market has monopoly tendencies. There is not enough political power within individual municipalities to take action to break these tendencies and introduce a regime of innovative developments. For instance, the other municipalities in the Stockholm region more commonly use waste-generated gas plants and biofuels in their energy systems than the City of Stockholm does. Therefore, proposals to create scale economies using a new common feedstock would create a tension between different parts of the region and between the com-panies that control the networks and the plants that feed those networks.

Garbage collection systems

The growth of the Stockholm region will cause a continued increase in the volume of waste to be recycled, deposited, or combusted. The Stockholm region does not yet have a fully thought-through strategy for garbage collection and waste separation. Waste separation is still voluntary, and organic waste cannot yet be handled effectively by the City of Stockholm. There are large differences between the City of Stockholm and the surrounding municipalities, which are partly a consequence of the large number of actors in the market for waste handling. Here again, there is a conflict between joint and individual solutions, and this may well be exacerbated in the future, for instance as regards water and wastewater handling in the coastal municipalities in particular.

The archipelago challenge

A major challenge for the development of the Stockholm County arises from the fact that the eastern part of the county borders the Baltic Sea and consists of more than 3,000 islands. A large number of areas originally built to accommodate recreational summerhouses are now permanently inhabited and are incorporated into the suburbs of Stockholm. Their physical character, dominated by outcropping bedrock, physiographical diversity, and attractive natural surroundings, makes infrastructure systems such as new road construction or district networks of water, wastewater, and energy difficult and costly to design and implement. Several of the municipalities east of Stockholm are facing a conflict between development of effective connections to major regional networks and the challenge of developing local solutions, especially for water and wastewater. Overextraction of water from private, drilled wells leads to groundwater salinization; and ineffective sanitary systems lead to nutrient leakage into the Baltic Sea (Olofsson and Rönkä, 2007). Innovative technical solutions – both local and subregional – must in many cases be implemented to allow housing development in these areas.

Telecommunications systems

The challenges surrounding the role of telecommunications in the future sustainability of the Stockholm region are multifaceted. From an infrastructure systems perspective, one long-term issue relates to the wiring of the building stock. The penetration of ICT solutions into the transportation system, including the vehicle fleet, has already started and is proceeding rapidly. Cloud computing and the internet have a bearing on emergent digital lifestyles, work–life structure, business opportunities, and e-democracy. For the moment, there is genuine uncertainty about future developments, which makes it particularly useful to reduce quasi-static uncertainty through focused research and work with scenario methods to handle remaining broad or genuine uncertainty.

Examples of research challenges in the field are, for instance, how to design advanced integrated travel planners that both provide efficient decision support for travelers and function as a tool for local and regional planning for sustainable development. Another field is data-driven sustainability, exploring innovative information technologies for creating, sharing, remixing, and visualizing sustainability information. One area with significant potential for sustainability impacts relates to the question of what circumstances will encourage organizations to use videoconferencing and other forms of mediated meetings that can be viable alternatives to face-to-face meetings requiring physical travel.

It is important to continue studying social practices involving ICT as a potential *engine of transition*, whereby changed practices can affect the sustainability impact of companies and organizations, as well as the visions, goals, policies, and concepts for sustainability. High on the research agenda is the deepening of our current understanding of the role of telecommunications for working life in a future sustainable Stockholm region, and the exploration of how the innovative application of ICT can contribute to more energy-efficient transport habits.

The complexities of sustainable transport infrastructure

Stockholm's transport system for both road and rail is considered by many to be in need of major investment in terms of both new links and refurbishment. This has given rise to major investment projects such as the Western Bypass, the City Track Tunnel, the Eastern Link, and also the first attempts to reduce traffic growth through traffic management using a congestion toll system and increased parking fees. A threat that needs to be taken seriously is the deterioration of air quality due to nanoparticles released from vehicles and roads. There is a chance that some remedy can be found to address this issue, at least in terms of the particulate emissions from vehicles, for instance through the introduction of electric cars and other electric vehicles, including bicycles. Some further examples are given in what follows.

Figure 5.4 shows the development of one basic transport system indicator over recent decades. It is evident that the growth of the region has been associated with a steady decline in the share of public transport in total transport volume. Over the past 40 years, Stockholm has made relatively modest new investments in both road and other automobile-related infrastructure and in the public transport system. The figure shows that the combination of total growth and growth distribution across locations has implied a move away from sustainability goals. The example shows the importance of traffic management measures to promote sustainability and infrastructure supply measures, namely the Stockholm congestion charge and the Stockholm subway system.

Congested roads seem to be an unavoidable characteristic of large cities. Transport economists and planners have often suggested that road pricing is an appropriate and effective instrument in an overall policy to relieve

PUBLIC TRANSPORT SHARE 1973 – 2009 (index 1973 = 100)

Figure 5.4 Public transport's share of total personal travel in the Stockholm region, 1973–2009

congestion (see, for instance, Rietveld and Bruinsma, 1998). Politicians and the public at large have, however, usually been quite skeptical. In Mattsson (2008) and Börjesson *et al.* (2012a), three *ex ante* (predictive) studies of transport and location effects of alternative road pricing systems are presented and compared together with one *ex post* (empirical) study of the implemented congestion pricing system in Stockholm. Alternative models estimated with different data sets are applied to calculate the effects. The first two studies deal with the effects on the traffic pattern of a zone-based and a distance-based road pricing system for the Stockholm area, respectively. In the third study, location effects are also included in an analysis of optimal congestion charges in a stylized symmetric city adjusted to resemble Stockholm.

All three *ex ante* studies indicate a substantial reduction in vehicle distance traveled as a result of imposing a congestion charge. For the zone-based system, traffic volumes in the inner city of Stockholm are predicted to decrease by 30 percent for charged hours at a charge level equivalent to 3 Swedish kronor (SEK) per kilometer (about €0.35 per kilometer). For the distance-based system, traffic volumes in the inner city are predicted to be reduced by 35 percent and 19 percent at charge levels of 4 SEK and 2 SEK per kilometer (about €0.50 and €0.25 per kilometer) during peak and office hours, respectively. For the case of optimal congestion pricing, the reduction is 25 percent at an average charge level of 2 SEK per kilometer. Additional effects predicted in the first study are that speed might increase on inner-city roads and arterials by around 20 percent. On the other hand, accessibility to activities in the rest of the city would be reduced significantly. Most affected would be the relation between inner northern and inner southern suburbs, for which a 30 percent reduction in the number of vehicle trips is predicted. In spite of quite substantial transport effects, the location effects (changes in settlement and workplace location patterns) are predicted to be very limited.

Börjesson *et al.* (2012a) conclude in their empirical *ex post* study that the traffic reduction effect of the Stockholm congestion pricing system has been quite substantial and that it has in fact increased during the pricing system's first five years of operation. The single most reliable traffic volume indicator, the number of vehicles that pass the cordon toll perimeter around the inner city, indicates a stable reduction of about 20 percent. However, when one also considers changes in external factors such as employment, fuel price, and car ownership, the study indicates that the reduction effect has most likely increased during the period. This is in line with theory predicting larger effects over the long term than in the short term.[1]

Börjesson *et al.* (2012b) perform a cost–benefit analysis of the Stockholm subway system as a whole, applying current Swedish methodology, models, and established national guidelines. The study forecasts the full social costs and benefits of the whole subway system as it existed in 2006, compared to the counterfactual situation (if the subway had never been built). The year 1956, prior to the major extensions of the subway networks in the 1960s and 1970s, is taken as the starting point for the calculations. Suppose that in 1956 the transport planners determining the appropriate level of investments in the subway system had had full information about land use in 2006. How would they have calculated the net present benefit to cost ratio of the 1956 investments?

Different assumptions are used in the no-subway investment scenario. In one analysis, it is assumed that no other public transport infrastructure has been built. In another no-investment scenario, the assumption is that the tram system that ran along the same corridors that the subway uses now had remained in operation. The analysis also varies the construction cost assumptions. The cost–benefit analysis is carried out using the actual construction cost from the 1950s and an estimate of what the cost would have been in 2006. The analysis also estimates the long-term effects on the labor market in terms of income growth and land-use patterns (which are not included in standard cost–benefit analyses).

The analysis shows a positive outcome for the subway scenario, and hence concludes that, in retrospect, the Stockholm subway system was a socioeconomically efficient investment. The analysis concerns the current subway system and gives no indication of the socioeconomic benefits of a possible expansion of the subway from the present time onwards.

Although the analysis shows that the subway was socioeconomically beneficial to build in the 1950s, some weaknesses of the methodology are also discussed. More specifically, the benefits of urban investment are typically underestimated in ordinary cost–benefit analyses. Such (under-estimated) benefits may, for example, include increases in regional labor market productivity because of increased commuter benefits, or improved matching in the labor market because the subway is widely used for commuting. The analysis demonstrates that the largest benefit of the Stockholm subway is its high capacity. Mere savings in travel time resulting from using the subway compared to the bus or tram are rather small.

Much of the current land-use pattern of Stockholm has been planned in conjunction with the building of the subway system or following its corridor structure. Börjesson *et al.* (2012b) also simulate how the Stockholm land-use pattern might have developed had the subway not materialized. The current system produces roughly the same socioeconomic benefits when one assumes the simulated land-use pattern as it produces given the existing land-use patterns that have adapted to the subway system. This is because the benefit of high-capacity public transport infrastructure is generally greater in cities with a dense settlement pattern.

The result of the land-use simulation indicates, hence, that Stockholm has developed into a sparser and less dense region than the one that market forces would have produced in the absence of the subway system. A consequence of the investment in the subway has been that new settlements were planned far from the city center. However, there is a high demand to live in the inner city, and higher density could have been the resulting land-use pattern if the development of the city were governed by market conditions (in the absence of the subway system); see Figure 5.5.

This analysis demonstrates that the practice of planning new settlements along rail corridors extending far out from the city center increases travel distances and even energy consumption compared to denser land-use patterns. Settlements along the subway corridors far from the center thus seem to have contributed to increasing trip distances, and have weakened the competitiveness of bicycles and other means of public transport. An important conclusion is therefore that it might have been better not to extend the subway lines quite as far out from the center.

Figure 5.5 Map of forecast land-use pattern, 2006, if the subway system had not been built (dark gray), compared to the actual land use pattern (light gray)

5.4 SUSTAINABLE INFRASTRUCTURE SCENARIOS FOR STOCKHOLM

This section uses a system-level analysis to discuss the relationships between infrastructure planning actions and market outcomes, using different assumptions concerning the development of external driving factors for the Stockholm region.

Figure 5.6 forms a starting point for the discussion of the concept "sustainable infrastructure." It is based on the so-called Brotchie triangle, which illustrates the relationship between urban form and individual commuting behavior, as well as how changes in technology affect these relationships (Brotchie, 1984; see also extensions in Newman and Kenworthy, 1998; Wegener and Fürst, 1999). The figure also illustrates the connection between mobility behavior (the ways in which people use the transport network) and the network structure for various spatial settlement and workplace patterns.

A given city could be characterized by a point anywhere within the triangle. The *x*-axis denotes the dispersal of urban activities – for instance, mean travel distance to employment from the center of the region. The *y*-axis shows the degree of spatial interaction, which could for example be measured as total or mean travel to work. The A value of zero along the *x*-axis indicates a perfectly monocentric city, and a value of 1 represents a fully dispersed pattern of activities. Point A, then, represents a situation in which all urban activities are centralized (everyone commutes into the central city), point B a situation in which people choose their jobs without regard to job location in relation to where they live, and point C a situation in which everyone lives and works locally. A dispersed urban settlement structure combined with a large spread in the mobility behavior of households and firms can give rise to a large volume of total kilometers traveled in the short and medium term.

Point D shows where a hypothetical city could be within the triangle. If we can estimate the position of a city in the triangle, the triangle construct also helps us understand how the city system would change as a result of various types of construction (which would shift D left or right) or of reorganization or moves (which would shift D up or down).

In a highly condensed format, Figure 5.6 thus describes the relationship between, on the one hand, the three factors of settlement patterns, workplace locations, and transport network morphology and, on the other hand, resulting sustainability performance, whereby the total volume of travel (which affects fuel use and climate change) can be seen as an indicator of sustainability. The Brotchie triangle illustrates the observation that if all workplaces are at the center of the region, it will be favorable (from a sustainability perspective) to develop a basically radial network. If everybody is able to find a job near their home, in the dispersed pattern of workplaces, where they are spread in proportion to the dwellings – that is, there is a common regional labor market – it will be less necessary to have an integrated infrastructure network. On the other hand, if workplaces are dispersed in proportion to the dwellings, and labor market specialization and mobility behavior are such that no local options are sought, it is better to develop a fully integrated transport system to boost accessibility.

The figure also illustrates how the interaction between the cost of travel and the level of dispersion affects the overall travel pattern (the vertical arrows inside the triangle). Higher transport costs for a given urban structure lead to a travel pattern closer to the lower boundary of the triangle, and vice versa. Consequently, if transport costs are high, a dispersed urban structure would minimize the volume of travel (vehicle kilometers traveled), while for low transport costs a dense structure would be travel-minimizing.

Lundqvist (2003) has performed calculations concerning the position of the Stockholm region in the Brotchie triangle. This analysis places the Stockholm region in the northeastern corner of the Brotchie triangle and concludes that successive regional development plans, combined with the planning actions by the municipalities and the public transport providers, have moved Stockholm's point D even further toward the many-to-many corner (see also Wegener and Fürst, 1999).

The above analysis suggests some important policy-relevant conclusions. The first observation is that a varied supply of transport opportunities

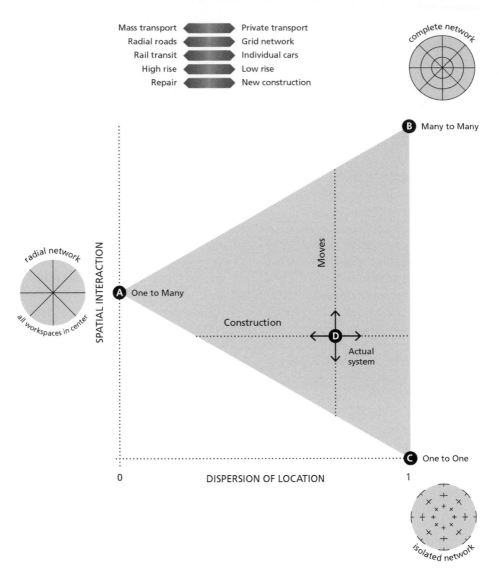

Figure 5.6 The interaction between urban structure, user behavior, and welfare development

at different regional scales is crucial in order to find a reasonable balance in the trade-off between sustainable development and urban growth. The composition of supply and the behavior of rapid transit providers will affect the long-term competitiveness of urban regions in light of uncertainties regarding the private and social costs of transport supply. A varied transport system will substantially increase the resilience of the region. This needs to be complemented by an urban structure characterized by regional subcenters with functional integration in land use in the context of a wider urban area with functionally specialized housing areas.

Mixed land uses at different scales can increase access to local amenities and thus decrease the total volume of travel. On the other hand, total travel volume could increase if travel costs remain low and the land-use pattern reflects increasing specialization. Telecommunications will likely not have a single directed impact toward decreased mobility, and there may be indirect effects of a different nature. For example, the so-called rebound effects described by Mokhtarian (1998) imply that advances in telecommunications that make for telecommuting and other ostensibly travel-reducing effects may in fact lead to more recreational travel or to more energy use being transferred to the lighting and heating of homes. In the long run, rebound effects will emerge through induced mobility accruing from the extension of opportunities to travel within the range defined through telecommunications. Rebound effects will accrue through the propensity of firms and households to use increased accessibility offered by telecommunications to travel more – a powerful mechanism that was convincingly presented by Hägerstrand (1973).

The scenarios described in Figure 5.7 illustrate some of the trade-offs between integration and isolation in the infrastructure and land-use

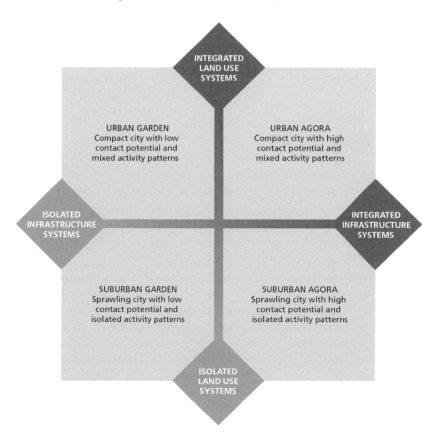

Figure 5.7 Four Stockholm region scenarios in the nexus between integration and isolation

system in the Stockholm region. Four scenarios have been constructed (see also Gullberg *et al.*, 2009). The URBAN AGORA scenario represents a situation where the region has become denser and infrastructure systems have become more integrated, accommodating a city in which interactions among people and firms are easy (a high contact potential) and interacting activity clusters are present (mixed-activity patterns).

The SUBURBAN GARDEN scenario, by contrast, represents a region sprawled into dispersed activity clusters with both low average density and low density in the suburban parts of the city. The transport and utility networks are developed along the same lines, with preference given to isolated local systems, each one in and of itself oriented toward the use of sustainable technologies in energy supply, water, waste disposal, and telecommunications.

The URBAN GARDEN and SUBURBAN AGORA scenarios are cross-combinations of the two extremes. In the URBAN GARDEN scenario, the city comprises isolated, mixed-activity areas with individual tailored solutions and little contact among neighborhoods. The SUBURBAN AGORA scenario builds on the idea of decentralized concentration, whereby settlements are dispersed across the regional territory, while the infrastructure systems are integrated within each settlement. A suburban settlement landscape with localized district heating systems will be a typical outcome in this scenario.

5.5 CONCLUDING DISCUSSION

The analysis presented in this chapter has sketched some development possibilities in the direction of sustainability originating from the current infrastructure systems of the Stockholm region. These actions might include introducing new systems, but might also be taken within systems themselves. New solutions may also include organizational and technical innovations. Even limited-scale demonstration effects can create an impact on sustainability goal fulfillment at the local, regional, and national level.

As several other chapters in this book suggest, a number of current infrastructure projects in Stockholm have components that can be exported as expressions of this paradigm. These results suggest that the mutual gains from international trade also apply to the markets for innovations, knowledge exchange, and demonstration projects. The currently ongoing project of applying the urban greening criteria developed by the OECD (Stockholm, 2012) to the Stockholm region is an interesting reflection of this trend. In particular, the OECD report notes the value of demonstration projects in supporting growth that is both economically and environmentally sustainable, by providing a showcase for innovative "green" technologies and firms.

Actions might also be related to changing the relationship between the infrastructure systems and the urban settlement systems at large, and the urban activity systems among firms, households, and public agencies. Infrastructure systems may be ostensibly sustainable within the system but differ in the ways they relate to and affect the sustainability of other regional systems. For example, transport infrastructure has strong locational properties, as it affects the location choices of households and firms. There might therefore be feedback effects in the form of changes to parts of the transport system that can both strengthen and weaken synergies among modes. An example is provided by the ridership reaction from investment in public transportation, and in bicycle track systems. Bicycle users are more likely to be former public transport users than former car users. Alternatively, persons who have been daily public transport commuters may select a more distant workplace and use the car to get there.

The situation is somewhat simpler with regard to other infrastructure systems such as energy, sewerage, and waste. If they are made more sustainable internally, but maintain the same service level, the feedback issues might be less pronounced. Of course, efficiency in energy distribution through district heat, for instance, necessitates densification at both local and regional level. As was mentioned earlier, substantial enhancements in the direction of sustainability are possible even without extending the systems. Investing in new capacity in new locations will often give rise to citizen opposition and raise questions with respect to inter-municipal cooperation, questions that require complex negotiations of issues in order to be resolved.

Economic policy instruments in the urban systems field can have a strong impact on goal fulfillment in the area of sustainability. Distributional effects are generic, and will in most cases give rise to blocking coalitions among both political groupings and enterprise networks. The effects may also peter out over time if economic instruments are not combined with proper institutions that guide the process of their planning and implementation and are not streamlined with supply actions. These instruments will have to be designed to secure stated and revealed freedom of action for firm cultures and lifestyles. For instance, the success of the Stockholm congestion toll system rests, to a considerable degree, on the choice of toll collection technology that does not require commuters to stop at toll cordons and collects road charges using a single monthly electronic bill.

Urban municipal actors are important as learning organizations in close cooperation with their citizens in regard to education, information, e-democracy, and marketing. It is also important that they have superior competence in understanding the shaping of the demand for urban amenities. Such proactive behavior can be as important as the decision to produce urban amenities within their own organizations.

Different municipal roles must be clearly discernible from each other, and effective solutions will differ among regions of different size and with different regional incomes. In the current context of Stockholm, which is characterized by increased competition among private firms wishing to become service providers in a deregulated market, municipal strategic planning competence must be strong. The pace and character of development raises demands for a new mode of thinking when it comes to the competence of local civil servants. This implies a need for the creation of new knowledge networks and increased efforts toward lifelong learning.

For an urban municipality, it might be just as effective to attain sustainability goals by forming strategic alliances with other municipalities, and agencies at the regional level, as by acting locally within the home territory. The temptation to be a free rider in an integrated region can be strong, but is generally of non-lasting value. This holds in particular for the classic cooperation game between central city governments and suburban municipalities in the Stockholm region. Comparative experience reported recently in research at the European level shows that overriding urban coalitions are scarce but effective in securing sustainable development (see Hårsman and Rader Olsson, 2002).

Most of the technical potential to shift Stockholm's infrastructure systems toward sustainability currently lies in the introduction of information and communication technology both for information and citizen influence and for economic efficiency. Properly used, this technology can push all parts of the Stockholm region in the sustainability direction at the same time as it can reduce the per capita production cost for service production. On the other hand, it can also lead to irreversible damage in the form of urban sprawl and exacerbated segregation patterns. Sustainable development therefore calls for a rethinking of the responsibilities of the classical technical utilities. Should privatization proceed at the risk of replacing public monopolies with private ones? How are we to conceptualize the notion of sustainable infrastructure?

Much has been said about the benefits of partnership between the private and public sectors. It must be acknowledged, however, that the public sector has been the most interested party when it comes to partnerships in planning and building activities. Moreover, the importance of and potential in creating incentives within both households and firms that support sustainability should not be underestimated. The introduction of car pools, car-sharing clubs, and new urban tram systems has, for example, been quite successful. Large firms, among which the municipality as an employer is one, can also be powerful actors. Finally, there are hopeful signs that demographic and cultural changes may mean that Stockholmers exhibit different preferences. A young generation of entrepreneurs and individuals in the Stockholm region may well have new attitudes toward family formation, social competence, mobility patterns, and environmental solidarity.

The Stockholm region has good potential to take significant steps toward sustainability. Regional development planning has long had a strong position in the region. Therefore, there are great expectations associated with the work to be started in the coming years on the new regional development plan, building on the most recent regional planning process, *RUFS* (Stockholm County Council, 2010). The new plan will be the result of a lengthy planning process during which new perspectives are formulated and new in-depth studies performed. This new plan, which is to be completed in the next five years or so, can provide a breakthrough for the sustainability of the Stockholm region. The role of Greater Stockholm as a contact-intensive region will have to be especially carefully analyzed. It is time for an initiative for the long-term development of the Stockholm region which assumes the mantle of the historically ambitious planning traditions of the region.

NOTE

1   Editor's note: Chapter 7 of this volume draws a somewhat different conclusion from the experience of the congestion charging scheme, suggesting that long-term effects may be smaller than short-term effects as commuters accept congestion charges as part of their regular transportation budget.

REFERENCES

Andersson, Å. E. and Strömqvist, U. (1987) *The Future of the C-society*, Stockholm: Prisma (in Swedish).

Andersson, Å E., Snickars, F., Törnqvist, G., and Öberg, S. (1984) *Regional Variety for National Benefit*, Stockholm: Liber (in Swedish).

Berdica, K. (2002) "An introduction to road vulnerability: What has been done, is done and should be done," *Transport Policy*, 9(2): 117–127.

Berdica, K. and Mattsson, L. G. (2007) "Vulnerability: A model-based case study of the road network in Stockholm," in A. T. Murray and T. H. Grubesic (eds.) *Critical Infrastructure: Reliability and Vulnerability*, Berlin: Springer, pp. 81–106.

Bettencourt, L., Lobo, J., Helbing, D., Kühnert, C., and West, G. (2007) "Growth, innovation, scaling, and the pace of life in cities," *Proceedings of the National Academy of Sciences*, 104(17): 7301–7306.

Börjesson, M., Eliasson, J., Beser Hugosson, M., and Brundell-Freij, K. (2012a) "The Stockholm congestion charges – 5 years on: Effects, acceptability and lessons learnt," *Transport Policy*, 12: 1–12.

Börjesson, M., Jonsson, D., and Lundberg, M. (2012b) *The Long-Term Benefits of Public Transport: The Case of the Stockholm Subway System*, Report 2012:5, Stockholm: Expert Group on Public Economics, Ministry of Economic Affairs, Stockholm (in Swedish).

Brotchie, J. F. (1984) "Technological change and urban form," *Environment and Planning A*, 16: 583–596.

Gullberg, A., Höjer, M., and Pettersson, R. (2009) *Images of the Future City*, Stockholm: Symposion, Brutus Östling (in Swedish)

Graham, S. and Marvin, S. (2001) *Splintering Urbanism: Networked Infrastructures, Technological Mobilities, and the Urban Condition*, London: Routledge.

Hägerstrand, T. (1973) "The domain of human geography," in R. J. Chorley (ed.) *Directions in Geography*, London: Methuen.

Hårsman, B. and Rader Olsson, A. (2002) "The Stockholm region: Metropolitan governance and spatial policy," in W. Salet, A. Thornley, and A. Kreukels (eds.)

Metropolitan Governance and Spatial Planning: Comparative Case Studies of European City-Regions, London: Spon Press.

Hudson, W. R., Haas, R., and Uddin, W. (1997) *Infrastructure Management*, New York: McGraw-Hill.

Jenelius, E., Petersen, T., and Mattsson, L.-G. (2006) "Importance and exposure in road network vulnerability analysis," *Transportation Research Part A*, 40: 537–560.

Johansson, B. and Snickars, F. (1992) *Infrastructure: The Building Sector in the Knowledge Society*, Stockholm: Swedish Council for Building Research (in Swedish).

Lundqvist, L. (2003) "Land-use and travel behaviour: A survey of some analysis and policy perspectives," *European Journal of Transport and Infrastructure Research*, 3: 299–313.

Mattsson, L.-G. (2008) "Road pricing: Consequences for traffic, congestion and location," in C. Jensen-Butler, B. Sloth, M. M. Larsen, B. Madsen, and O. A. Nielsen (eds.) *Road Pricing, the Economy and the Environment*, Berlin: Springer, pp. 29–48.

Mokhtarian, P. (1998) "A synthetic approach to estimating the impacts of telecommuting on travel," *Urban Studies*, 35(2): 215–241.

Newman, P. and Kenworthy, J. (1998) *Sustainability and Cities: Overcoming Automobile Dependence*, Washington, DC: Island Press.

Olofsson, B. and Rönkä, E. (2007) "Small-scale water supply in urban hard rock areas," in E. Plaza (ed.) *Urban Water Management: Is There a Best Practice?* Urban Management Guidebook 1, Baltic University Urban Forum (BUUF), Uppsala: Baltic University Press, pp. 21–26.

Olsson, K. (2002) "Planning for the preservation of the cultural built heritage," in F. Snickars, B. Olerup, and L. O. Persson (eds.) *Reshaping Regional Planning: A Northern Perspective*, Aldershot, UK: Ashgate.

Rietveld, P. and Bruinsma, F. (1998) *Is Transport Infrastructure Effective? Transport Infrastructure and Accessibility: Impacts on the Space Economy*, New York: Springer.

Stockholm (2012) "OECD Green Cities – Stockholm background report," draft report, City of Stockholm.

Stockholm County Council (2010) *RUFS: Regional Development Plan for the Stockholm Region*, Executive Summary, Stockholm County Council.

Törnqvist, G. (2011) *The Geography of Creativity*, Cheltenham, UK: Edward Elgar.

Wegener, M. and Fürst, F. (1999) "Land-use transport interaction: State of the art," Institute of Spatial Planning, University of Dortmund.

CHAPTER 6
THE ECONOMICS OF GREEN BUILDINGS

Hans Lind, Magnus Bonde, and Agnieszka Zalejska-Jonsson

6.1 INTRODUCTION

IN THE PAST DECADE, *green buildings* have undergone dramatic development in Stockholm – and indeed throughout Sweden. Construction materials and technologies have developed quickly, and from an investor perspective green building has gone from being an almost unknown concept among developers and investors to being something that is increasingly seen as a necessity. To produce buildings that are not classified in some way as "green" is not an option in Stockholm's current real estate market. This is a process that is almost completely driven by the market: buildings are built green primarily because it is profitable to build them green, not because the government demands or subsidizes them.

In this chapter, the label *green buildings* is used with respect both to buildings that are ranked highly in environmental ranking systems (see the next section) and to buildings where the focus is on reducing the amount of energy needed to heat them – such as passive houses. It was not until 2008 that the property sector in Stockholm started to take a broader interest in green buildings. The first report to present a systematic discussion of the economic aspects of green buildings was published by Sweden's Royal Institute of Technology (KTH) in 2009 (Bonde *et al.*, 2009). The report covers environmental classification systems, price and rent effects, green leases, and valuation methods. This led to a large body of postgraduate research covering a broad spectrum of issues related to the economics of green buildings, with a focus on energy efficiency.

KTH was involved from the beginning in the development of an environmental classification system adapted to the Swedish situation (see Malmqvist *et al.*, 2010), as will be described in more in detail in section 6.2. Two examples of newly produced green buildings will be presented in section 6.3. Section 6.4 will describe a number of studies on the economics of building green, and section 6.5 will discuss what is happening with the existing stock, focusing on the development of the Stockholm market and providing an overview, based on Swedish studies, of our current understanding of the economics of green buildings.

6.2 ENVIRONMENTAL RATING TOOLS IN SWEDEN

The concept of a green building in the Swedish context sometimes refers to a building's energy performance and sometimes to broader aspects that are included in environmental classifications, such as LEED or the Swedish system. A number of systems are in use in Stockholm, and Sweden as a whole: the main standard ratings for buildings used are LEED from the United States, BREEAM from the United Kingdom, EU Green Building, and the Swedish system *Miljöbyggnad* (Environmental Building). The more general "Swan" label (related to the EU Ecolabel) is also used.

EU Green Building focuses only on energy efficiency, and the general view in Sweden is that it is very easy – even too easy – to be awarded this classification. In the case of renovations, it is often easy to reduce energy use by 25 percent. Producing new buildings that use 25 percent less energy than the official national standards is also rather easy. International building construction firms in Sweden have generally chosen either LEED (e.g. Skanska) or BREEAM (e.g. NCC), but so far only a small number of buildings have been classified with these labels. The Swedish Green Building Council, a non-profit organization that promotes certification systems and awareness, reports that about 250 EU Green Building- and Environmental Building-certified buildings, somewhat over 50 LEED certified buildings, and some-what below 50 BREEAM certified buildings have been built in Sweden.

The Swedish system *Miljöbyggnad* was developed by the academic community (including KTH) and representatives of the building and property sector. The public sector in Sweden in particular has started to use the *Miljöbyggnad* system on a rather large scale. The Swedish system focuses on energy use, environmentally friendly materials, and indoor climate, but is somewhat narrower than LEED and BREEAM. The main reason that Sweden developed its own environmental classification system was that the inter-national systems were seen as too complex and costly; certifying according to *Miljöbyggnad* is less expensive than LEED and BREEAM certifications. Further, the existing international standards did not completely align with those aspects of green building that are in focus in Sweden. For example, as Sweden has a large supply of fresh water, water consumption is considered less critical than it is in other contexts. Further, Swedish development actors wanted a system that focused on the building only itself, rather than on issues such as the trade-offs between brownfield and greenfield development, or transportation issues, arguing that many of these aspects are not under the control of the developer. Another argument was that the weighting system used in systems like LEED and BREEAM is based on cumulative points, while the Swedish *Miljöbyggnad* system classifies buildings on the basis of the aspect, category, or dimension where the building has lowest points, meaning that low points with respect to one dimension cannot be compensated for by high points in relation to another. The building thus has to be green in all respects in order to be certified. This, however, does create some difficulties when evaluating renovations, because even if some dimensions are greatly improved, the rating remains unchanged as long as the lowest-dimension scores are unaffected (Brown *et al.*, 2011).

The Nordic "Swan" Ecolabel is a voluntary assessment system for products and services. A company applying for a Nordic Ecolabel for small houses, apartment buildings, or preschool buildings must fulfill requirements relating to the construction process, the choice of materials, and energy performance. Buildings are evaluated on their environmental impact through lifecycle analysis, with the aim of minimizing hazardous substances in construction material, maximizing energy efficiency, and promoting the sustainable disposal of construction waste.

6.3 COMMERCIAL AND RESIDENTIAL GREEN BUILDINGS IN STOCKHOLM

Commercial

In 2006, Vasakronan (a major commercial real estate owner in Sweden) decided to rebuild the office building Pennfäktaren 11, located in central Stockholm. The building was constructed in 1976 but was not optimal for more modern office activities. This building had been rented by the Swedish postal service and by a police department, but these tenants had moved to other locations. The site itself had also formerly been the headquarters for a newspaper; in 1971, that building had been torn down.

Vasakronan outlined three different options for the refurbishment of the building: "light" renovation, more profound renovation, or a major renovation or rebuilding. It decided to proceed with the third option. The project did not have any specific environmental goals from the outset, but as refurbishment planning continued, environmental issues assumed a higher profile. When the project entered an active phase in 2009, it had the character of a "green" rebuilding project.

The renovation faced some technical challenges, foremost related to changing the load-bearing framework. The building had formerly relied upon a quite complicated concrete structure, but this was replaced by a load-bearing steel frame, allowing the ceiling height of the offices to be raised considerably.

Figure 6.1 Pennfäktaren 11, a commercial building in the Stockholm central business district
Vasakronan

The refurbished building offers eight stories of office space (about 15,500 m² of rentable area), a street-level story with restaurants and shops, and a garage below street level. In order to reduce the building's energy usage, it was refitted with several green features, such as:

- a demand-controlled ventilation system and lighting system that saves energy by adapting to the number of people in the building;
- a sedum roof ("green roof") that improves insulation and absorbs air pollutants;
- energy-efficient windows;
- solar collectors and solar cells: solar collectors are used to heat the restaurants' tap water and provide comfort cooling during the summer; the solar cells generate electricity, which is used in the building.

These improvements have resulted in considerable reductions in energy usage, from 250 kWh/m² yearly before the renovation to 100 kWh/m² yearly after the refurbishment of the building. In addition, the basement is equipped with a recycling station, and Vasakronan also plans to sign a "green lease" with each of the building's tenants, a commonly determined agreement to capitalize on opportunities and share the savings associated with energy use reductions.

Because of its strong environmental performance, the building has been pre-certified at a Gold level in accordance with the LEED Coor and Shell scheme,[1] a globally accepted environmental assessment system for evaluating a building's environmental performance. In addition, the building has been EU Green Building-certified. This system only considers the energy performance of the building; in order to be certified, a building has to have its energy usage lowered by 25 percent (in the case of renovations) or use 25 percent less than the national building code requires (in the case of new construction).

The building has also been one of five finalists in two building competitions, *Årets Stockholms byggnad* (the Stockholm Award for Building of the Year) and *ROT-priset* (the ROT Award). Both of these competitions are awards for retrofits of the existing building stock.

Residential

In 2008, construction began on a rental apartment estate in Farsta, in the southern part of Stockholm. Two years later, four perfectly ordinary-looking, multifamily buildings were ready to be occupied. Yet the housing estate, called *Blå Jungfrun* (the Blue Maiden), is far from an ordinary residential estate. Owned by municipal housing company Svenska Bostäder and built by Skanska, these are some of the few apartment buildings in Sweden that fulfill rigorous passive house standards, referred to as the Swedish "FEBY" (*Forum för Energieffektiva Byggnader*) standard 2007:1.

Interestingly, initial plans for the Blue Maiden housing estate project did not include building passive houses or low-energy buildings. However, early collaboration between Svenska Bostäder and Skanska led to a commitment to work as partners and created space for innovative solutions. The result is four buildings comprising 97 rental apartments for which annual energy and hot water consumption is lower than 50 kWh/m². This means that average energy consumption is calculated to be 60 percent lower than expected energy performance according to current building regulations (which is 110 kWh/m² per year in the Stockholm region).

To achieve substantial energy reductions, buildings must be airtight and have a very well insulated envelope, highly energy-efficient windows, and controlled ventilation with a heat recovery system. Blue Maiden buildings are constructed using a VST system – that is, using prefabricated elements, with Styrofoam (*cellplast*) insulation and highly energy-efficient windows with a *U*-value of 0.9 W/(m² K). Blower door tests were conducted and showed excellent results, achieving values as low as 0.05 l/s m², surpassing the highly energy-efficient passive house standard requirement of 0.3 l/s m² at ±50 Pa.

Figure 6.2 The Blå Jungfrun ("Blue Maiden") housing estate in Farsta

Jan Sernesjö

The consequence of an airtight building envelope, good insulation, and passive design is that a building's energy requirements become very low, so low in fact that there is no need to install a traditional heating system. Instead, each building is equipped with a central mechanical heat-exchange ventilation system that supplies heated air to each apartment. Simulations show that in well-insulated, airtight buildings, air heating is sufficient to deliver good indoor thermal comfort. However, since winter temperatures in Sweden can be very low, the Blue Maiden apartments include small electric heaters in each apartment. These heaters are equipped with thermostats and timers, allowing occupants to control usage. In order to help tenants control and minimize total consumption, each apartment is equipped with a so-called SBox, a display that allows occupants to follow their energy consumption on a daily or monthly basis. Electricity use, as well as cold and hot water consumption, is measured and benchmarked against calculated and predicted values. The results are presented in kilowatt-hours or monetary units, giving tenants the opportunity to influence their behavior and anticipate the expected charge that will appear on their utility bill.

Measured values for energy and water consumption in the Blue Maiden development are significantly lower than those of other buildings in the existing housing stock of Svenska Bostäder, representing as much as a 65 percent reduction, and actual energy performance in the first year of operation differs by only 8 percent from expected values. Post-occupancy surveys conducted in 2011 indicated that the vast majority of tenants were pleased with their apartments and proud to live in an environmentally friendly building. Since 2010, other green residential buildings have been built in the Stockholm region. Svenska Bostäder has built another passive housing complex in Annedal, on the border of Stockholm and the municipality of Sundbyberg, and is in the progress of building its third multi-apartment building using passive house technologies, in the Stockholm suburb of Blackeberg.

Green apartments are available not only on the rental market but also on the ownership-condominium market. In the western part of Stockholm, NCC has built two condominium buildings that meet passive house standards and were certified according to the FEBY standard. The buildings received certification at the silver level within the Swedish *Miljöbyggnad* environmental assessment system. Since 2010, *Miljöbyggnad* has registered and certificated a number of other buildings in Sweden, some of them in the Stockholm region. Another condominium building designed to have very low environmental and climate impact is Veidekke's *TellHus*. The building is the first multi-apartment building to be awarded the Nordic "Swan" Ecolabel.

Residential buildings in Sweden may also be certified according to LEED building assessment standards, and LEED has been used by Skanska in a handful of such projects. Since the assessment scheme originated in the United States, there are some potential difficulties in adopting it in a Scandinavian context (for example, as has already been mentioned, water consumption matters less in Sweden). However, those barriers can be overcome, and the proof is that over 200 apartment condominium buildings in the City of Stockholm are LEED-certified.

Green development and innovation diffusion in construction

Looking back at the past decade and following the development of more energy-efficient buildings, we can easily notice that as little as ten years ago, *green* or *highly energy-efficient* products and materials were rather uncommon in construction. Nowadays, the same products are in standard use, and are often recommended by building regulations.

One of the most important components in building design is the choice of windows. The thermal protection requirement for windows has improved significantly. Heat losses from windows used in 2000, and even in 2010, were twice or even three times higher than heat losses from windows commonly used in Sweden today. The average *U*-value for windows used in Sweden is about 1.1–1.3 W/(m^2 K), and more developers are installing windows with values recommended for passive house construction (i.e. 0.9 W/(m² K)) in their conventional building projects.

Buildings with high energy efficiency must be airtight and have very good insulation. The gap between *green* and *conventional* buildings used to be significant with respect to these aspects. However, a decade later the building envelope of almost all newly constructed buildings is more airtight and better insulated than ever. The main factors behind this development are increased accuracy during construction, technical improvement in wall construction, and minimized thermal bridges. Ten years ago, a very well-insulated building envelope meant that walls were very thick, mainly because of the addition of many layers of insulation. This is not necessarily the case today. The newest generation of bricks combines traditional clay and insulation in one block, which means that a wall built entirely of such bricks can have a *U*-value as low as 0.12 W/(m² K), and no additional layer of insulation is needed. This value can be compared with the passive house recommended level of heat transfer for walls, which is 0.15 W/(m² K). Wall thickness can be also reduced by using new-generation isolation panels, for example vacuum or PIR elements.

A healthy indoor environment with fresh, clean air is of paramount importance to any building. Technical developments in ventilation systems have also been significant. High-quality ventilation with heat recovery systems helps save more energy than is required to operate them. Ventilation systems using heat exchangers reduce energy demand because they make use of energy that was already in the building, taking heat from exhaust air and passing it through cold, fresh air. The newest heat exchangers on the market have very high energy efficiency, even up to 90 percent.

Promoting material safety, health, and the minimum use of hazardous substances are all actions that have also received special attention within the construction industry over the past decade. Detailed information about the materials used in construction products is now provided, and the product's health and environmental impact assessment and classification are made publicly available. There are a few standard methods and data sources in use in product evaluation and certification in Sweden, including *SundaHus* and *Byggvarudeklaration* (or BASTA). Information about a product's health and environmental impact is very important, particularly in cases where the building's performance and environmental impact are being assessed, for example by *Miljöbyggnad* or the Nordic "Swan" Ecolabel.

6.4 ECONOMIC STUDIES OF GREEN BUILDINGS IN SWEDEN: NEW CONSTRUCTION

How profitable is it to build green, or to improve the environmental standards of a building? Another way to put the question is: What is the level of greenness that we can expect private actors to implement, assuming that they will act in their own best interest? Studies related to this have been carried out all over the world, the most famous of them being that by Eichholtz *et al.* (2010), who found a clear correlation between environmental qualities, rent levels, and property value. Here the focus will be on studies undertaken in Sweden, and especially in relation to the Stockholm market. The framework used in the analysis below starts from a standard cash-flow analysis, wherein the profitability of an investment depends on investment cost, rent levels, operating costs, the rate of return demanded, and exit value.

Construction cost

Private construction companies do not publish their construction costs, but some more indirect studies have estimated the difference in construction costs between ordinary and green buildings. Zalejska-Jonsson (2011) undertook interviews and conducted a questionnaire with clients and contractors who had experience in building green, and found that the average construction cost increase for a residential passive house was around 5 percent. A similar result was found for the construction cost of a house-standard day-care center. An interesting observation noted in this study was that private actors estimated the difference in construction costs between traditional and passive houses to be lower than the difference estimated by public-sector clients.

Rental income and tenant satisfaction

Several studies indicate that most tenants who move into green buildings, whether commercial or residential, choose the building for reasons other than the environmental profile of the building (see, for example, van der Schaaf and Sandgärde, 2008; and, for studies of the commercial sector, Diaz-Jernberg and Ytterfors, 2011). In recent years, demand for rental housing has been higher than its supply, particularly in areas where rent control regulation keeps rents well below market rates. Those seeking rental units thus tend to value the general attractiveness of the area and the location more than how green the building is. Location was also the most important factor for commercial tenants.

Zalejska-Jonsson (2011) found that most households that moved into green residential areas chose those areas for reasons other than their environmental profile, indicating that a lack of willingness to pay extra for renting in such a building existed. However, even in the case of tenants who had moved into a green building more or less by chance, many reported feeling proud to live in a green building and that they had become more environmentally conscious. This indicates that in the future they may have a higher willingness to pay for renting a unit in a green building. The same pattern was also found for tenants in commercial buildings who started to use the environmental profile of the building in their marketing efforts; this may make it more difficult for them to move into a non-green building in the future. A comparative study of tenants in green buildings (in this case referring to buildings built with passive building techniques) and traditional new residential buildings found that even though in some cases the passive house tenants used extra heating in the winter, the tenants in the passive buildings were on average somewhat more satisfied with the indoor climate than those living in more traditional residential buildings.

Operating and maintenance costs

The most obvious advantage of green buildings is lower energy cost. Zalejska-Jonsson (2011) found that people in the sector estimated energy costs to constitute on average 20 percent of operating costs. An open question, however, has been whether or not the more advanced technologies used, especially in the ventilation systems, create special problems both initially and in the longer term if the equipment does not work as planned. Interviews with property managers with experience of both green and traditional buildings suggest that this has not been a problem in reality. Initial technical problems were as common in traditional buildings as in the green buildings.

Risk, borrowing cost, and rate of return demanded

The value of a green building and the profitability of building green will also be affected by how banks and other actors on the financial market evaluate them. For example, the value of a building increases if it is possible to finance it at a lower interest rate. Jaffee and Wallace (2009) argue that a green building is less risky than an ordinary building, as the cash flow will be less sensitive to changes in energy prices. Therefore, a rational bank should be willing to finance a green building at lower interest rates. Empirical studies in Sweden do not, so far, support this view. Ohlsson and Retelius (2010) find that banks still focus primarily on cash flow and risk to cash flow, and think that it is too early to consider giving special treatment to loans on green buildings. First, it would have to be shown that the risk actually is lower. Nevertheless, the focus on cash flow indicates that energy savings and reduced operating costs are considered to be the most important aspects from the bank's perspective; as the improved cash flow strengthens the company, it is easier to get a loan, all else being equal. This also means that it could be possible to borrow somewhat more money to build a green building or get a loan on somewhat more favorable terms, even if the reason is not directly related to the greenness but rather to the anticipation of improved cash flow.

Investors' views on green building

There are no special government funds that provide incentives to construct green buildings in Sweden. On the other hand, in recent years pension funds have demonstrated more interest in investing in real estate as the expected rate of return on other assets has fallen. No green profile can be detected in these investments, however. One study focusing on investors (Reuterskiöld and Fröberg, 2010) found that there were investors interested in investing in "non-green" buildings because it can be profitable to upgrade them into greener buildings. The picture of investors was generally rather similar to that of the banks: the investor might be willing to pay more for a green building if the net operating income is higher because of lower energy costs. The investors also believed that demand for green buildings would increase in the future, and that this increased demand would make such buildings more interesting to them.

Sundbom (2011) found that building non-green commercial build-ings was no longer an option for the large property companies in Stockholm, at least in central locations. Companies believed that it would be difficult to find tenants in new office complexes if the building were not environmentally certified. Building green is now the norm for this type of building and it would be a risk to the company's brand to produce a new building with lower environmental standards than other recently produced buildings in the same area; it would signal that the property owner did not care about the environment.

Greenness and property value

Property values reflect net operating income, risk, and expected future developments. The first Swedish study on how valuers look at green building was published in 2008 by Bonde *et al.*; the main finding of the study was that at that time the only aspect that mattered was the way in which environ-mental characteristics affected net operating income. The valuers did not believe that owners of green buildings could charge higher rents, but as the operating cost was lower, the net operating income would be higher, and therefore the value would be higher. These studies did not investigate whether or not the value increase would be enough to compensate for higher production cost. More recent studies, including one by Kuiken (2009), support this view. The number of green buildings is still too low to make it possible to directly measure their effect on rents, vacancies, and prices. The reduction in operating cost and the improved net operating income remain the factors affecting value.

Using a survey to collect environmental information about recently sold single-family houses, Mandell and Wilhelmsson (2011) found that a number of environmentally related factors, such as the type of ventilation and heating system used, had a price effect. Most of these factors also reduced the operating costs of the buildings, so the result of their study is in line with the hypothesis suggested above: property values are only affected by the greenness of the building if they reflect a reduction in operating costs. There is no extra increase in property "just" because it is green.

It is interesting to note that when the international construction company Skanska decided to increase its cooperation with KTH, it was in the area of green buildings, and one question that was especially in focus in that collaboration was how to value green buildings in a way that reflects the full breadth of positive environmental effects that they offer.

6.5 GREENING THE EXISTING BUILDING STOCK

Commercial and public-sector buildings

Several studies show a number of improvements in energy performance in the existing building stock. Maneekum (2009), for example, looks at shopping centers, and Backteman (2011) studies office buildings in suburban locations. Typically, there are a number of "low-hanging fruits" that offer large savings from relatively inexpensive and/or simple initiatives, such as using more energy-efficient lights and motion-sensitive lighting, adjusting installations, installing taps that reduce hot water consumption, and (more generally) adjusting the service level to reflect actual need. For example, ventilation and heating can be reduced when the space is not used.

Working systematically with these low-hanging fruits can reduce energy consumption by up to 30 percent and can be highly profitable, with payback periods under three years. Even if there are no systematic studies covering the whole stock, our impression is that most professional property owners are working with this issue today in Stockholm. The group that seems to be lagging is, unsurprisingly, small private landlords, who are not strict profit-maximizers but are rather happy with their current situation: inertia is strong within this group. The recent requirement to produce an energy declaration is one way to make this group more aware of their situation and about what can be done to improve the environmental performance of their buildings. The Swedish Property Owners Federation has also produced information for these property owners in order to make them aware of the profit opportunities present in energy savings.

The first sectors that worked systematically with energy savings were the public authorities charged with providing special-purpose buildings such as schools and hospitals. Over the past ten years, a number of reports have been produced by the Swedish Association of Local Governments and Regions, which collects experience and provides advice. A special prize is given to the authority that has been most successful, and links to good examples are available on its website.

Residential buildings

Between 1963 and 1973, 1 million homes (houses and apartments) were built across Sweden as part of the so-called Million Homes Program, and many of these units are now in need of renovation. This situation is seen by many as a window of opportunity, since renovation of technical systems can be combined with measures to reduce energy consumption. The owners of multifamily rental housing, which dominates the Million Homes Program, respond very differently to these opportunities. On the basis of an interview study and a questionnaire, Högberg *et al.* (2010) and Högberg (2011) identify four types of property owners: The *strict profit maximizers* only invest in initiatives with a payback period of around three years. The *little extra companies* are willing to reduce the rate of return demanded by a modest margin. Two types of *ambitious companies*, driven by an ambitious board or ambitious managers, were also identified in the studies. Private companies could be found in the first two groups, while municipal housing companies could be found in all four groups. Municipal housing companies generally have boards of directors that include local political representatives.

A crucial question is how profitable energy savings are. Is there really an energy efficiency gap, with several profitable measures that companies for one reason or another do not undertake? Another controversial question has also been how to calculate profitability. Should one look at a whole package of measures together, such that the most profitable measures essentially subsidize the less profitable, or undertake a more traditional marginal analysis, taking the most profitable first and then asking whether an extra step is profitable given the savings already made on the first measures? The choice of evaluation method can have a considerable effect on what is seen as profitable.

Lind and Högberg (2011) argue that, using standard methods for calculating profitability, it would be possible to reduce energy consumption, given current prices, by around 30 percent. It was further argued that companies which did more than this often were motivated by factors other than economic gains. The idea of an energy efficiency gap was questioned. Mahapatra *et al.* (2012) present results from a project focusing on owners of single-family houses and find that two factors hinder this group from making profitable energy-saving investments: a lack of knowledge, and difficulties in identifying serious contractors. We can, however, see that investments in alternative energy sources, such as heat pumps, have been rather high even if investment in such technology is still lower than investment in reducing energy use in other ways, for example by using extra insulation and changing ventilation systems. They argue that in the long term, the owner-occupied residential market, including multifamily condominiums, could be a greater hindrance to energy savings at the societal level than the rental housing stock.

Green leases, individual metering, and contractual innovation

A question of special interest, especially in the office market, is how responsibilities should be allocated between tenant and landlord. Here we can find large differences among countries. Sweden represents one extreme, as the typical contract includes almost everything in the base rent, including heating and operation and maintenance costs. The other extreme is common in countries such as the United Kingdom, where the tenant has a long-term contract and is responsible for operation and maintenance. The cost of heating is generally paid directly by the tenant. The property owner is then more of a financial investor and the rent paid primarily covers the capital costs of the building. The underlying problem in a case like this is one of split incentives. Energy use in a building depends on both the actions of the landlord and the actions of the tenant. If the tenant pays the heating costs, then the property owner has no incentive to improve things like insulation or the ventilation system to recover more energy. On the other hand, if the landlord pays the heating cost, then the tenant has no incentive to economize on heating.

Green leases started to be discussed within the Swedish commercial market in late 2009 (see Ödman, 2010). Vasakronan, the biggest commercial property owner in Sweden, was one of the first to introduce green leases as an alternative for their tenants. This picked up rather quickly during 2010 (see Grahn and Jonsson, 2010). In a typical green lease negotiation, the landlord offers to map tenants' energy use and then a partnering structure is created whereby landlord and tenant meet regularly to evaluate energy consumption and plan possible measures. Even if it is not possible for technical reasons to measure exactly how much the specific tenant saves, there is a bonus for the tenant when energy use is reduced. Experience from these leases shows, however, that in economic terms the gain accruing to tenants from reduced energy use was rather small.

The tendency in Sweden is toward individual metering and charging of energy use, especially in new houses, where metering technology can be introduced at a rather low cost. This is most common for hot water, but individual metering of ambient air heating is also becoming more common. If a house is new, or newly renovated, it could be argued that the landlord has already made improvements to the building, and what remains is to provide effective energy-saving incentives to the tenant. As the building ages, however, individual metering and other methods that require the tenant to pay the energy cost can reduce the incentive for the landlord to improve the property. The dominating contract form in Sweden is one where the rent is all-inclusive and heating is included. Even if this reduces the incentives for the tenant, our hypothesis is that in a longer perspective, incentives remain for the property owner to make continuous energy-saving improvements. In order to improve incentives for tenants, a partnering structure like that introduced by Vasakronan may be a better option than individual metering and charging.

Bonde (2012) has studied the introduction of a green lease in a more complicated case with an institutional property owner, a property management company with a long-term contract with the property owner, and a large tenant renting the whole building. Even if the initial incentives were far from optimal, it turned out to be impossible to agree on the terms for a new greener lease, for example with respect to how much the rent should be reduced if the tenant took more responsibility for the heating costs. From an efficiency perspective, it would in cases like this probably be better if the user of the building also owns the building, thus making coordination of measures easier. Creating strong incentives for energy efficiency investments seems to be more difficult both in the rental office stock and when management is outsourced, especially when lease contracts are short (three to five years).

Green leases are not the only type of contractual innovation that can improve the environmental performance of buildings. Integrating construction and operations maintenance in a single contract could give life-cycle cost aspects a more prominent role. Different types of public–private partnership (PPP) have been under discussion in Sweden, but have not been implemented at a larger scale. PPP has been questioned from a theoretical perspective related to differences in writing long-term contracts that create the right incentives (see, for example, Lind and Borg, 2010) but, more importantly, the current Swedish center-right government has generally opposed PPP projects on the grounds that they promote an expanded public sector.

Energy Performance Contracting, a performance-based procurement method whereby reduced energy costs resulting from the installation in turn pay for the cost of investment, can also be found in the Stockholm market, but not to a large extent. On the other hand, the concept of green procurement is commonly applied in Stockholm in many other contexts (see, for example, Faith-Ell, 2005, for examples of how these procurements are carried out). The main idea is that the client prescribes a number of environmental qualities that the contractor must fulfill.

6.6 CONCLUDING DISCUSSION

During the past five years, there has been a dramatic increase in interest in green buildings and in energy-saving measures. Most of the larger real estate companies today have management executives who specialize in environmental issues. The remaining question is thus whether the ambitious environmental goals will be reached.

If the focus is on energy consumption per square meter, new construction seems to be the least problematic area: technologies are readily available, cost-competitive, and reliable. From a broader environmental perspective, urbanization and a higher level of housing construction will, however, have negative environmental effects in terms of increases in both material use and emissions during the production process (see Toller *et al.*, 2010). As has already been mentioned, it seems possible to reach a reduction in energy consumption of around 30 percent in the existing stock in a profitable way and without any special measures, but this is far from the long-term goal of a 50 percent reduction. So what can be done?

The economists' standard answer to the question of how to reduce energy consumption is, of course, higher energy prices. In Sweden, special subsidies or tax deductions for energy improvements have been discussed. These can be found in many countries, but for the moment Sweden has no such subsidies. Another policy is to promote technological innovations that would reduce costs for energy improvements. There is also evidence that in some cases, such as the private single-family housing market, a lack of information might be an important obstacle; therefore, developing the system of energy declarations might have some effect. If property owners were required to evaluate the environmental performance of their buildings more regularly, and also were given continuous feedback such as suggestions for improvements, energy saving might move higher up the agenda. Knowing what alternatives exist, and how to go about implementing them, can reduce the transaction costs that property owners face when making energy-saving investment decisions.

If we look at the development of green buildings in Stockholm from a broader perspective, one reflection is that this development could be seen as "typically Swedish," in the sense that the work is systematic and carried out on a rather large scale through stepwise technical improvement, but also in that it lacks more spectacular technical and architectural features. An advantage of this may, however, be that the proliferation of green buildings is not so dependent on fashion, external events, and specific government policies and that we can expect continuous improvements in the coming years.

NOTE

1   LEED Coor and Shell is a scheme meant to address sustainability in the initial design/development of the construction. The scheme assesses base building elements (e.g. structure, envelope, HVAC-system).

REFERENCES

Backteman, O. (2011) Environmental investments in commercial buildings (in Swedish), Master's thesis, Division of Building and Real Estate Economics, KTH, Stockholm.

Bonde, M. (2012) "Difficulties in changing existing leases – one explanation of the 'energy paradox'?" *Journal of Corporate Real Estate*, 14(1): 63–76.

Bonde, M., Lind, H., and Lundström, S. (2009) "How to value energy-efficient and environmentally friendly buildings" (in Swedish), Division of Building and Real Estate Economics, KTH, Stockholm.

Brown, N., Bai, W., Björk, F., Malmqvist, T., and Molinari, M. (2011) "Sustainability assessment of renovation for increased end-use energy efficiency for multi-family buildings in Sweden," in *Proceedings of the Sixth World Sustainable Building Conference*, SB11 Helsinki, 18–22 October.

Diaz-Jernberg, J. and Ytterfors, S. (2011) "Different industries' approaches in green leases" (in Swedish), Candidate thesis, Division of Building and Real Estate Economics, KTH, Stockholm.

Eichholtz, P., Kok, N., and Quigley, J. M. (2010) "Doing well by doing good? Green office buildings," *American Economic Review*, 100: 2492–2509.

Faith-Ell, C. (2005) "The application of environmental requirements in procurement of road maintenance in Sweden," Doctoral dissertation, Department of Land and Water Resources Engineering, KTH, Stockholm.

Grahn, A. and Jonsson, N. (2010) "Green leases" (in Swedish), Master's thesis, Division of Building and Real Estate Economics, KTH, Stockholm.

Högberg, L. (2011) "Incentives for energy efficiency measures in post-war multi-family dwellings," Licentiate thesis, Division of Building and Real Estate Economics, KTH, Stockholm.

Högberg, L., Lind, H., and Grange, K. (2009) "Incentives for improving energy efficiency when renovating large-scale housing estates: A case study of the Swedish Million Homes programme," *Sustainability*, 1(4): 1349–1365.

Jaffee, D. and Wallace, N. (2009) "Market mechanisms for financing green real estate investments," Fisher Center Working Papers, Fisher Center for Real Estate and Urban Economics, Institute of Business and Economic Research, University of California, Berkeley.

Kuiken, H. (2009) "Valuation of sustainable developed real estate: A closer look at factors used when valuing green buildings," Master's thesis, Division of Building and Real Estate Economics, KTH, Stockholm.

Lind, H. and Borg, L. (2010) "Service-led construction – is it really the future?" *Construction Management and Economics*, 28: 1145–1153.

Lind, H. and Högberg, L. (2011) "Incentives for energy efficiency measures in the housing stock from the 60s and 70s" (in Swedish), Division of Building and Real Estate Economics, KTH, Stockholm.

Mahapatra, K., Gustavsson, L., Haavik, T., Aabrekk, S., Svendsen, S., Tommerup, H., Ala-Juusela, M. and Paiho, S. (2012) "Business models for full service energy efficient renovation of single family houses in Nordic countries," paper given at the International Conference on Applied Energy, Suzhou, China, 5–8 July.

Malmqvist, T., Glaumann, M., Svenfelt, Å., Carlsson, P.-O., Erlandsson, M., Andersson, J., Wintzell, H., Finnveden, G., Lindholm, T., and Malmström, T.-G. (2010) "A Swedish environmental rating tool for buildings," *Energy*, 36 (4): 1893–1899.

Mandell, S. and Wilhelmsson, M. (2011) "Willingness to pay for sustainable housing," *Journal of Housing Research*, 20: 35–51.

Maneekum, F. (2009) "Green shopping centers: A survey of Stockholm's shopping center market and its environmental commitment," Master's thesis, Division of Building and Real Estate Economics, KTH, Stockholm.

Ödman, L. (2010) "Green leases" (in Swedish), Division of Building and Real Estate Economics, KTH, Stockholm.

Ohlsson, S. and Retelius, A. (2010) "Granting of credits to green buildings: Will favorable loan conditions exist?" (in Swedish), Candidate thesis, Division of Building and Real Estate Economics, KTH, Stockholm.

Reuterskiöld, A. and Fröberg, L. (2010) "Investors' view on environmentally certified properties" (in Swedish), Master's thesis, Division of Building and Real Estate Economics, KTH, Stockholm.

Sundbom, D. (2011) "Green building incentives: A strategic outlook," Master's thesis, Division of Building and Real Estate Economics, KTH, Stockholm.

Toller, S., Wadeskog, A., Finnveden, G., Malmqvist, T., and Carlsson, A. (2010) "Environmental impacts of the Swedish building and real estate management sectors," *Journal of Industrial Ecology*, 15 (3): 394–404.

van der Schaaf, K. and Sandgärde, M. (2008) "Environmental demands by tenants when seeking new premises in commercial properties" (in Swedish), Division of Building and Real Estate Economics, KTH, Stockholm.

Zalejska-Jonsson, A. (2011) "Low-energy residential buildings: Evaluation from investor and tenant perspectives," Licentiate thesis, Division of Building and Real Estate Economics, KTH, Stockholm.

CHAPTER 7
PERFORMING SUSTAINABILITY
INSTITUTIONS, INERTIA, AND THE PRACTICES OF EVERYDAY LIFE

*Ebba Högström, Josefin Wangel,
and Greger Henriksson*

7.1 INTRODUCTION

> "People will get help. When one buys a technical device, one gets a heavy handbook, something you seldom get when moving into an apartment," says environmental and sustainability strategist Tomas Gustafsson, when being interviewed about the new urban development project the Royal Seaport.
>
> Besides learning about the technical solutions – how to use a garbage disposal unit, how to measure how much water you use and how much waste you throw away (it is going to be charged individually) – the Residence School will be about what food you eat and how to buy eco-labeled goods."
>
> (Excerpt from an article in the Swedish national
daily newspaper Dagens Nyheter (Tottmar, 2010))

WHEN *DAGENS NYHETER*, Sweden's biggest newspaper, reported that a Stockholm urban development project planned to educate future residents about how to live an environmentally sound and healthy everyday life, the article caused an upsurge of protest and debate. The project under discussion was the Stockholm Royal Seaport, a high-profile sustainable urban redevelopment project and the new flagship in Sweden's and Stockholm's green-tech armada. To many, the project represented an unwanted revival of the social engineering type of public health promotional programs popular in Sweden during the 1930s and 1940s.[1] Others went so far as to deride the project as an example of "eco-fascism." However, at the very same time as debate on the return of social engineering raged, the glass and steel building GlashusEtt (Glass House One) continued its practice of educating and promoting sustainable lifestyles, as it had been doing for 11 years. GlashusEtt is the environmental information center of the Hammarby Sjöstad project, a renowned example of sustainable urban development which attracts study delegations from all over the world. The mission of GlashusEtt is to educate the neighborhood's inhabitants and visitors on environmental matters; in comparison to the Royal Seaport project, this mission was accepted without protest or questioning. Indeed, the Hammarby Sjöstad project had also seen its share of conflicts regarding environmental measures, but here these took the shape of power struggles over issues such as the number of parking spaces, the placement of windows, and the preservation of an endangered beetle.

A stone's throw away from GlashusEtt, car commuter trips originating in the southern suburbs are registered by congestion charging devices. These are mounted above each road leading into Stockholm's inner city; they not only register vehicles' passage but also inform the driver about the charge (in Swedish kronor) using large signs displaying "10 SEK," "15 SEK," or "20 SEK" (about €1.15, €1.75, and €2.30, respectively). For drivers, the congestion charges mean that every passage into the inner city during weekday working hours has become significantly more expensive.

The sustainability of a city is fundamentally dependent on its inhabitants and their activities. The recognition of this is no novelty: as early as 1272, King Edward I of England banned the burning of sea coal in London, and in the 1380s the first sanitary laws were passed in Paris and Cambridge. Besides such legislative measures, the palette of policy instruments in addressing sustainability also includes economic incentives (including both whip/push and carrot/pull types), information campaigns, and alterations of urban space. These policy instruments are generally directed either at inhabitants themselves or at the producers or providers of the goods and services consumed (e.g. through the adoption of a green building code) – in which case they are aimed at altering the environmental burden while not necessarily changing people's ways of life.[2]

The deployment of policy instruments is one of the main reasons why Stockholm, and Sweden in general, has been comparatively successful in promoting and implementing sustainable development. As compared to other, similar high-income countries, Sweden's environmental footprint is rather small – still substantial and unsustainable, but nevertheless smaller.[3] The possibility of using policy instruments, however, must not be taken for granted; using such instruments should rather be seen as a result of Sweden being historically characterized by a governance model with a strong state and strong local governments. The latter are especially powerful when it comes to the field of urban planning: Swedish municipalities not only have a monopoly on land-use planning, but also have their own financial resources because they have direct income tax authority.

This chapter discusses three recent cases wherein the City of Stockholm has, through the use of different kinds of policy instruments, tried to make the lifestyles of Stockholmers more sustainable. First, we look at two proclaimed "sustainable urban development projects" and measures related to them. We then turn our attention to the introduction of congestion charging in Stockholm. Our focus is on the institutional aspects of urban governance for sustainable development. However, rather than only looking at institutions in terms of governmental bodies and the way these are organized, which would be the typical political science perspective, we explore and describe institutions from a sociomaterial perspective. This means that we also assign *agency* – the power to influence – to the material aspects of politics, policies, and everyday life. We argue that materialities have the power to mediate discourses and practices by creating and promoting both inertia and change in urban systems. The agency and mediating capacity of these materialities are known, even if only in a tacit sense, to planners and policymakers – otherwise there would be little sense in deploying them.

Thus, in order to understand institutions we need to look closely into how they are formed and constituted, considering material and spatial aspects (e.g. things, objects, machines, devices, bodies, rooms, architecture, and buildings), discursive and normative aspects (rules, regulations, visions, ideas, and moral implications), and the everyday practices performed by the people affected by and/or using them. Furthermore, since our cases all describe top-down policy instruments, it is also crucial to include the interplay of scales, for example the interplay between the macro level institutions of governance and planning and the micro-level practices of citizens' everyday life, so as not to get stuck in seeing these as dichotomies.

This theoretical point of departure will be developed further in the following section. Besides the sociomaterial constitution of institutions, this section also introduces related notions of *path dependency* and *discourse*. Thereafter, our three recent cases from Stockholm will be presented. These case studies were selected in order to exemplify the ways in which a sociomaterial and discursive perspective can enable an analysis of the agency of institutions in transition toward urban sustainable development.

7.2 INSTITUTIONS, PATH DEPENDENCY, AND DISCOURSES

From a sociomaterial perspective, the world appears a seamless web in which material and social aspects are interwoven and mutually structuring. Neither the material nor the social can therefore be changed without also taking account of the other. This interconnectedness takes place at all levels of society. At a micro scale, a sociomaterial perspective would address the workings of everyday life through the practices performed to support our needs and wants; these workings would be seen as being imbedded in, and interacting with, macro-scaled structures. The macro scale addresses policy, legislation, and recommendations, but more implicit structures can be found here too: techno-political paradigms and regimes, and sociomaterial path dependency.

In policy and planning practice, there is, however, typically an imbalance, filled with contradictions, in how sociomateriality is approached. Material changes (e.g. changes in buildings, infrastructures, or technical systems) are typically depicted as rather unproblematic, since materialities are most often perceived as neutral and objective, in contrast to the value-laden social aspects of planning and policy making. However, with a socio-material starting point it becomes crucial to acknowledge that materialities are no less value-laden than the social; materiality encompasses policy, and is therefore entangled with discourses and power mechanisms. For instance, sustainable transport includes not only material changes (such as redesigning the transport infrastructure, or shifting to more energy-efficient cars and biofuels) but also mobility management schemes (such as congestion charges, or changing norms and habits). For example, even if biofuels were the only solution considered, for this to have an effect the car industry would need to be interested in producing such cars; and there would need to be car buyers willing to buy them, farmers willing to grow the biofuel crops, and fueling stations equipped to provide the fuel. All of these factors depend on a number of institutionalized sociomaterial aspects – rules, regulations, pro-cedures, legislations, norms, practices, experiences, values, and the supply of material space and devices (cars, biofuel gas stations) – as well as the way those aspects interact.

Material spaces and objects are not neutral. Their organization and design comes from somewhere: from an idea, an intention, a vision, or a norm. Spaces and objects are shaped within these discourses and with the intention of achieving certain outcomes. But as spaces and objects are established as material assemblages (by being built or being manufactured), this discursiveness often becomes less tangible; ideas and intentions that had informed the design become less obvious and the impression is that material space and objects act "on their own" to a certain extent, or merely function as some form of neutral material backdrop in front of which the influence of more apparent social, cultural, political, or economic forces plays out. Spaces and objects thus often quite covertly (and therefore perhaps all the more efficiently!) influence the practices and experiences of their users. Think, for example, of how walls and openings, as well as locks, influence your ability to move freely in a building. It is worth mentioning that the intentions that acted as a starting point for the design can be curtailed by material and spatial organization if the translation of the idea into physical form is not undertaken in an informed way. This means that the spatial and material organization need not always be supportive of initial ideas and visions; it can just as well be constraining and contradictory to them.

A sociomaterial approach such as actor–network theory (ANT) draws on assumptions of an entangled relation between the social and the material: within ANT, it is not meaningful to make any division between these two concepts. An ANT approach places focus on the performative capacity of space and objects, a capacity that is seen as an effect of networks of relations between human and non-human actors such as things, animals, or species (Law, 2002). Thus, from a sociomaterial perspective there are no predefined structures, and no distinction exists between the agency of humans and the agency of non-humans. Rather, they can be considered as equal *actants* in the establishment and upholding of network stability. The network's actants are not distinguishable; therefore, the network appears unified and is perceived as a taken-for-granted reality – that is, as a form of institution or, as Latour (2005) puts it, a "black box." This means that we can talk about, for example, psychiatry, economy, markets, or hospitals as pre-defined objects with stable borders. The effects of a stabilized network can therefore be seen where path dependency comes into play as it is difficult to change a stabilized network.

Institutions play an important role in determining which practices are awarded a space. For that reason, the design of institutions in terms of procedures and regulations, as well as material and spatial configurations, establishes important settings for path dependency. Path dependency does not mean that things do not change, but is rather a way to understand why change takes place in a certain direction. Path dependency is the occurrence of self-reinforcing or positive feedback processes that grow strong enough to prevent or obstruct any divergence from the status quo – that is, from the *path* (Pierson, 2000). This should not be misinterpreted as determinism, but rather seen as a sociomaterial process in which economic, material, and sociocultural factors combine to create a *logic of practice* that conceptually curtails our future choices (Kay, 2005). Path dependency is thus a way to explain why some alternative actions seem more or less appropriate than others and, by extension, why changing the course of societal development is such a difficult task to accomplish. This relationship between institutions and practices is not, however, one-way; practices also influence institutions. New practices and discourses can call for new institutions to be established, such as the emerging institutional support for urban agriculture that has recently grown out of an increased citizen engagement in this practice. To avoid becoming obsolete, authorities and organizations must be sensitive to changes in society and respond to these in a way that is aligned with those changes, or else manage to successfully take control over practices in terms of pointing out the "right way" to do things.

In this respect, material space and objects stand out as having especially inert capacities, owing to their physical and material character. Once a material space has been constructed, it is not easy to get rid of, to change, or to demolish. In this sense, material space frames path dependency; as the actor–network approach would put it, once the network is stabilized, it is often not perceived as changeable.[4] However, meanings also change over time, and new interpretations emerge that can change spatial experiences, even if the material configuration remains the same. Space has therefore both symbolic and material power. Discourses operate on a representational level and inform spatial visions, ideas, and intentions. Materiality, on the other hand, both enables and constrains activities and the use of physical space in terms of its performance: materiality and discourse *do* something, together with their users. Discourses are materialized in policies and institutions either as texts, procedures, routines, and buildings and are, as such, perceived as stabilized networks – as truths.

Institutions can be seen as a formalization of habits (Czarniawska, 2005). Before habits become rules (i.e. before they become institutionalized), they act as discursive objects in a struggle over meaning, over what is considered a problem, how this problem should be formulated, what is considered a feasible solution, and what is prioritized. A discursive struggle is thus a matter not so much of rational arguing as of arguing about what is considered rational. An institution could thus be considered to constitute a stabilized discursive practice (i.e. a stabilized network) or a formalization of habits into rules, norms, codes of conduct, and materialities (i.e. buildings, machines, weapons, pass cards, etc.). Recognizing path dependency to be the result of sociomaterial processes therefore calls for us to turn our focus toward the relationships between the constituents of user practices, material space and objects, and institutional procedures.

In the following sections, we unfold two stories where good intentions regarding the promotion of sustainable lifestyles did not work out as expected. We start with the information center, GlashusEtt, in Hammarby Sjöstad, which was supposed to act as an educational facility for sustainable living but instead ended up acting like an advertising tool for maintaining the good reputation of Hammarby Sjöstad as a sustainable city district. We then discuss a city district in the making, the Royal Seaport, where the intentions and methods for changing the future inhabitants' lifestyles were extensive but not anchored in the prevailing discourse about what local governments should and should not do. A materially spoken tiny thing such as an article in a newspaper could therefore if not overthrow, at least delay these plans.

7.3 FROM TEACHER TO PREACHER: THE STORY OF GLASHUSETT

In 2005, together with the energy company Fortum, the City of Stockholm released an advertisement for "a city district where some of the buildings are unusually intelligent" (Bylund, 2006, p. 137). The advertisement also included a picture of a building saying "Hi!" to a passerby walking a dog (Figure 7.1).

Det finns en stadsdel där vissa hus är ovanligt intelligenta.

McBride/Illustration Göran Uggla

I Hammarby Sjöstad ligger GlashusEtt. Det är ett "klokt" hus som är närmast själv-försörjande på energi. El och värme kommer till exempel från solceller på taket, bränsleceller och biogas. Till och med husets väggar används som solfångare. Förutom att titta på ny smart energiteknik kan du här också studera den miljövänliga kretsloppsmodellen. Lösningar som med all säkerhet kommer att användas i nya stadsdelar i framtiden.

Välkommen till vårt miljöinfocenter GlashusEtt i Hammarby Sjöstad, så får du veta mer!

 TILLSAMMANS FÖR EN HÅLLBAR ENERGIFÖRSÖRJNING I HAMMARBY SJÖSTAD

www.hammarbysjostad.se

Figure 7.1 Poster from GlashusEtt, the environmental information centre in Hammarby Sjöstad

Fortum

The city district is Hammarby Sjöstad and the building is GlashusEtt, the district's environmental information center. This section explores what more than "Hi!" the building GlashusEtt is saying, to whom, why, and how.

The story of the sustainable city project Hammarby Sjöstad and the building GlashusEtt begins in the mid-1990s when the City of Stockholm decided to make the already partly planned Hammarby Sjöstad project a central part of Stockholm's bid to host the 2004 Olympic Games (Green, 2006). This decision caused fundamental alterations to the formal institutions of the Hammarby Sjöstad project, through the local-area environmental program that was developed as a result (Svane *et al.*, 2011). The environmental program comprised three main parts: a number of environmental objectives guided by a "twice as good" environmental performance standard, the establishment of a project team to realize the project, and the setting up of an *environmental center* (Stockholms stad, 1997). Hammarby Sjöstad was intended to become an international role model not only for green construction but also for green lifestyles reflecting a sense of environmental responsibility. Information and pedagogy were tools for enhancing awareness of the correlations between lifestyle and environmental influence, this in order to make the residents of the area feel co-responsibility for their environmental influence and engagement in the area's environmental investments. Clear feedback showing the effects of the inhabitants' behavior was also to be given (Stockholms stad, 1997). GlashusEtt was aimed at facilitating the contributions of the inhabitants to the goals in the environmental program.

The center was to be established and run by the City of Stockholm, the Stockholm Water Company, and the energy company Fortum. Tengbom Architects were commissioned for the architectural design and at their website describe the building in highly technical and functional terms through descriptions of technical devices and systems for sewage, water, power, along with room specifications (reception and information desks) and building technologies (e.g. a sedum roof and solar cells). Hammarby Sjöstad's website (www.hammarbysjostad.se), edited by staff at GlashusEtt, mentions similar technologies but describes them as "eco-technologies that characterize the building."

The Hammarby Sjöstad website (Hammarby Sjöstad, 2012) does however also describe the building in another way:

> The residents' involvement is an important part of the environmental work. GlashusEtt provides advice on how to have less environmental impact and conserve resources among other things. It is also here that the residents can get [paper] bags for free for their biodegradable waste. In the entrance hall, you can see a model of the entire area and the eco-cycle is visualized through lights and buttons. The room is also used for different kinds of exhibitions that inform about current environmental issues. Those are in Swedish,

but can easily be understood by foreigners or explained to you by our staff.

GlashusEtt was inaugurated in 2002, its main task being to function as an environmental information center for the inhabitants of the area. In February 2012, the one hundred thousandth visitor was registered. This means that the center has had an average of 30 visitors per day. Today, the center has around 12,000 visitors a year. However, a closer look at these impressive numbers reveals that only 3,000 (25 percent) of the visitors are residents of the area. The other 9,000 are external visitors, of whom 7,000 come from abroad. This indicates that the center's original main target group has been replaced by a focus on informing visitors from outside the area, which of course has consequences for the extent to which more sustainable practices among the inhabitants can actually be promoted.

Using an actor–network theoretical approach, it is feasible to look upon GlashusEtt as an actor in the network *sustainable urban development*. But the actor GlashusEtt is also a network in itself, comprising an assemblage of actants – that is, all the humans and non-humans involved in its estab-lishment and maintenance, including the building and its eco-technology, the staff, the website, and the information material produced. As they, through assemblage, constitute an actor, they "speak with one voice" (Czarniawska and Hernes, 2005), and thus the building itself, the activities housed in it, and its stated mission are aligned with one another into an integrated object, a whole.

As an actor, GlashusEtt performs, or is evoked to perform, three somewhat different but still interconnected roles, with both discursive and material connotations. First, through its physical presence in Hammarby Sjöstad, GlashusEtt is a constant reminder to the area's inhabitants to take their environmental responsibility seriously and to "go green." The image it conveys is that of a sustainable city district with environmentally aware and responsible inhabitants. Second, it is also a powerful actor in the marketing and export of Swedish sustainable city models and green technology. In this sense, GlashusEtt not only performs its function as an information center to which delegates and business representatives can be taken, but also is represented and evoked in numerous information materials, articles in newspapers and trade papers, and television shows; information about it is printed, broadcast, and published online. Third, GlashusEtt is also an actor that is used within the City of Stockholm: first, in order to build momentum for the idea of developing the Royal Seaport as a sustainable city district, and second, to support the idea that this area too should have an information center. The discourses promoting a sustainable lifestyle are thus performed by GlashusEtt through the building as a material symbol of environmental responsibility, through its potential as a space for education, and by its many visual representations in the media and in information material.

Without going into any counterfactual reasoning about what would have been the case if GlashusEtt had not existed, it can still be concluded that the actual materiality of the building – the space and meeting place provided by it – facilitates a manifest non-linguistic representation of the discourse of sustainable urban development. Even though the original mission of GlashusEtt was to promote and support sustainable practices among the inhabitants in the area by acting like a teacher in the neighborhood, the building and its use have become more like a preacher of the Swedish sustainable city discourse. In this role, to judge by the number of visits and coverage in media and politics, the building has done a good job. But as a teacher, GlashusEtt has been less successful: the residents comprise but a small share of the total number of visitors, the energy use in buildings in Hammarby Sjöstad is about twice as high as the target figure, the decrease in water use is due to passive technologies only (low-flushing toilets, etc.), and the inhabitants do not source-separate their waste more, or use their car less, than the average inner-city Stockholmer. With this in mind, it is interesting to see how GlashusEtt is being used as an argument for locating another information center in the Royal Seaport, as well as how this planned center is also framed as a teacher and a facilitator of meetings.

7.4 IMPOSING RESPONSIBLE LIFESTYLES IN THE ROYAL SEAPORT: ECO-FASCISM OR INDIVIDUAL CHOICE?

Stockholm's new flagship for sustainable urban development, the Royal Seaport, is an urban development project in the making. Ideals of urbanity, sustainability, and healthy, livable environments are clearly formulated in the vision, which expresses intentions for the development to become a "Swedish sustainable city" (Stockholms stad, 2009). Social, ecological, and economic sustainability, as well as the "responsible lifestyles" of the citizens (i.e. both healthy and environmentally friendly), are emphasized in various aspects of the urban planning and design.

The Royal Seaport is also intended to represent and illustrate the sustainable city as a Swedish export good (Stockholms stad, 2009) and thus take over this role from Hammarby Sjöstad. The old brownfield area beside the Royal National City Park is planned to accommodate 10,000 new apartments and 30,000 workplaces. The overall objective of the development is the creation of a climate-adapted and fossil fuel-free urban district by 2030, in which carbon dioxide emissions are lower than 1.5 tons per person per year by 2020 (Stockholms stad, 2012). Clearly, there is need for more policy measures that extend beyond providing economic incentives for green technology and information if the sustainability challenges are to be met.

Figure 7.2 Perspective sketch of the envisioned environment at the Royal Seaport Beach Park
Stockholms stad, Andersson Jönsson Landskapsarkitekter

One stated aim in planning for this "sustainable city" is to foster sustainable lifestyles via changes in people's behavior. The idea of educating people to pursue more appropriate lifestyles is not a new one, but it seems that actions taken by public authorities to steer, shape, and change attitudes (and thus behaviors) are doomed to heavy critique in contemporary market-oriented society.

The specter of a totalitarian top-down imposition of lifestyles was invoked by the Swedish newspaper *Dagens Nyheter* in November 2010 with the following front-page headline in large type: "Promise to Live a Green Life – Get an Apartment." Accompanying the front-page article was a photo showing a woman in her thirties jogging on a leaf-strewn pavement. Adjacent to the pavement, a row of cars are parked; on the other side is a forest. In the caption to the image, the woman, Ulla Källström, states that she is eager to live in the new Royal Seaport area, where the inhabitants are supposed to live a green life and exercise regularly.

Ulla Källström jogs six times a week, recycles her garbage, generally buys ecological and fair trade food, and prefers to take public transportation. In the article, she represents the well-informed citizen who already lives a healthy and environmentally friendly life. She is convinced that we are all responsible for taking care of the environment and demands from the politicians to tell the inhabitants what is needed. In her opinion, the Royal Seaport is an experiment: a modern, socially fostered eco-village (which she approves of). The extent to which this outlook is emphasized in the marketing of the area, however, differs from what is stated in the newspaper coverage. The real estate developer JM is already selling apartments in the area but does not demand any commitment to sustainable lifestyles in its promotion material – rather, it stresses individual choice:

> Live life in a world-class environmental urban district. Just beside Östermalm and the Royal National City Park, the Royal Seaport is emerging, an attractive address for urban people who love nature. The district's stated environmental profile puts quality of life and sustainability in first place. Here you live on your own terms, at your own pace, and you can easily choose a conscious and sound lifestyle.
>
> (JM, 2012)

In this quotation, the desire to grab urbanity, nature, and environment at the same time, without imposing any predefined lifestyle but simply implicitly mentioning that the choice of a sustainable lifestyle is to be made individually, somewhat curtails any ambitions for greater agreement around reducing environmental impacts. This is an example of contemporary governance structures, in Foucault's (2010) terms "governmentality," whereby power is not imposed on you but rather aims at becoming internalized within your own will and thus becoming ostensibly executed as a "free" choice.

The official documents of the Royal Seaport formulate a series of objectives intended to secure a sustainable development; most of the focus areas of the documents are on technical systems: climate-adapted and green outdoor environment; water, waste, and energy recycling; energy; and environmentally efficient transport. Another heading addresses social practices ("Living and working in the area"), describing the Royal Seaport as a place for demonstrating solutions showing the consequences of our choices, and "how each of us can make a difference in the bigger picture" (Stockholms stad, 2012). Thus, everyone living and working in the Royal Seaport has a part to play in reducing energy consumption. For example, measures are planned that allow people to see how consumption and costs increase and decrease depending on individual resource use. It is stressed that participation and commitment to choices made by inhabitants and businesses will play a crucial role in the development of this eco-district.

Overall, there is an ambiguity between on the one hand clearly stated guidelines and regulations, and on the other a faith in the power of information and a reliance on a rational human capable of making "good" choices of their own free will. It should also be kept in mind that the area where people live can only form a part of the overall circumstances influencing their everyday life; in order to make good choices, inhabitants also have to overcome *outside* barriers, such as general market and policy conditions for jobs, goods, and services in society. The Royal Seaport wants to be an environmentally friendly, carbon-free, vibrant city district with a beneficial entrepreneurial climate, and at the same time be inhabited by people living healthy and environmentally friendly lives. This is supposed to be achieved by establishing a strong consensus culture. The City is to use its power as a landowner to formulate criteria framing agreements for land use and land development, thereby securing environmental objectives as well as the vision of a vibrant district. In attending to this task, the City acknowledges that "realizing the vision of a world-class environmental urban district is a huge amount of work and the City often acts to lay down requirements to drive development further" (Stockholms stad, 2012). These requirements cover, for example, the number of parking spaces, local electricity production targets, and rules to ensure that the domestic appliances installed (such as refrigerators and washing machines) have the highest energy efficiency ratings. It seems easier, however, to establish requirements for materialities that are measurable, such as energy efficiency and renewable energy solutions, than to impose restrictions on the everyday practices of the inhabitants.

As was noted earlier, another article in *Dagens Nyheter*, "Pressure on residents for healthy lifestyles", acted as a starting point for a debate on social engineering. The concept stems from the critique of the planning of the Swedish welfare state, which has been heavily criticized from the 1970s onwards (Hirdman, 1989). Historian Karin Johannisson (2010) makes this point in a follow-up article and notes similarities and differences between the traditional social engineering policies, such as Swedish public health programs and implementation efforts, and today's entrepreneurial and individualistic health projects. If the welfare state period was characterized by a strong state involvement in public health promotion, it is now outsourced to form part of a consumerist culture in which individual choice is the guiding star. The private sector is expected to achieve the goals broadly formulated by the public authorities. Personal health and the quest for an environmentally responsible lifestyle (i.e. a "sustainable lifestyle") are thus posited as an individual project and are promoted as such. "If you do not take care of your body, you do not take care of the environment" is the message communicated, further implying that if one does take care of one's own health, one does not need to engage in any further overall structural environmental action, thereby passivizing citizens against a wider collective political mobilization around the climate issue.

Overall, there is an emphasis being placed on voluntarism, by both private developers and politicians. A representative from the real estate developer JM, Gunnar Landing, says in the *Dagens Nyheter* article that JM does not demand that their customers are environmentally friendly; this has to be voluntary. In the same article, a member of the Social Democratic Party and of the City of Stockholm Traffic Council, Jan Valeskog, is quoted as being worried about the low number of per capita parking spaces planned in the Royal Seaport and fears that this will be perceived as a prohibition on families owning a car.

There have, however, been cases for which the fear that a measure would appear as a prohibition has been less of an issue, for example when the City of Stockholm introduced a system of congestion charges. The idea of using congestion charges to reduce car traffic and thus emissions caused considerable debate and criticism at first. However, once the charges had been levied for the trial period, they seemed to be remarkably easily accepted by the general public. One reason for this might be that after a period of time, the charges became internalized to the habit of driving a car in Stockholm, causing the economic incentive effect to become almost intangible.

7.5 THE (IN)VISIBLE WHIP: THE INTRODUCTION AND PERMANENT ESTABLISHMENT OF CONGESTION CHARGING

Policy instruments in the form of economic incentives directed at inhabitants are aimed at actually steering people's practices. Congestion charging means that many Stockholmers receive a bill every month, or make a direct debit payment, as a modest reminder of the societal consequences of their car

Figure 7.3 One of the access roads leading to the central parts of Stockholm, including the information and registration devices mounted above the charging cordon of the access road. To the right, partly hidden by a bush, is the EU sign for pay roads (coins in a hand), with a price list below

Stockholms Stadsmuseum Archive, Ramón Maldonado

trips. Driving into Stockholm, you pass devices mounted above the access roads to Stockholm's inner city. During rush hours and office hours, the devices register passage of vehicles, and the registered owners are charged at the end of each month.

After decades of futile attempts to introduce congestion charging in Stockholm, an agreement to make a full-scale trial was finally approved shortly after local and national elections in 2002. However, it would take a full season to determine the actual implementation of the policy, which meant that the Stockholm trial started on the cold and snowy day of 3 January 2006.[5] Media coverage preceding the introduction had been intense, and protests and problems were anticipated. In light of this, the introduction went rather smoothly. It was as though the snow muffled the noise and emotion so that the charges could quite easily slip into the city's real-world political arena.

The trial charges were levied from 3 January to 31 July 2006. The political objective behind the introduction of congestion charging was to control traffic by affecting citizens economically. Politicians wanted to influence a significant proportion of inhabitants to leave their car at home, switch to public transport, park and ride, stay home, skip a trip, or travel at another time. Pay stations (see Figure 7.3) consisted of electronic equipment, cameras, and signs located on the portals of all road entrances to Stockholm. Drivers were not required to stop when passing the pay stations. Each passage was registered automatically by cameras and also by transmission to a minuscule automatic transponder box mounted on the inside of the windshield which recorded all passages if the owner of the car wished to have the tax charged directly to his or her bank account. The transponder was free and not compulsory.

Traffic planning in Stockholm mainly takes place at a regional level. This means that 26 municipalities have to cooperate with each other, as well as with county and state authorities. Adding to this are business actors and economic interests. Into this mix of considerations and coordination, congestion charging suddenly landed on citizens, evaluators, journalists, researchers, and many other groups that might react with anticipation, curiosity, and skepticism. Where would this go?

Reactions to the Stockholm trial among residents in the region showed a wide variety of attitudes to the idea and practical ways of coping with the new situation (Henriksson, 2008). Here is one example:

> *Tina*: For me it has been a marked improvement. It takes me 25 minutes to drive to work instead of 45–50 minutes. So the reduction in congestion has made it even more attractive for me to travel by car to work: a paradox really. Just because I can afford to pay. Though most people might be able to afford to pay. So I'm a little surprised that traffic volume has decreased so much. But I can tell

you how I thought before I even came up with the idea that I would actually continue to drive and pay the tax. I had my mind set on using public transportation instead of my car. But at some point I said to my husband, "I'm concerned that it will take a very long time. How will it go? I often work late." And to make it home from Frihamnen [Tina's workplace] in the evening is not possible. "Yes, but it is simple to pay," he said then. "If you think it's worth getting to work and home quickly, then just pay the congestion charge." He was right, of course. But I was on another track at first.

Interviewer: But why were you on that other track to begin with?

Tina: I'm not sure. It was this general reluctance to pay for driving, I think. But that antipathy has disappeared since the effect has been so positive."

(Henriksson, 2008, p. 40)

Tina was not alone in marveling that the traffic dropped so significantly. However, along with the majority of motorists in the county, Tina initially expressed clear aversion to congestion charging. No wonder, since the trial was a political project and a new tax that affected many people's lives. The trial was a specific, but to some extent also flexible, constellation of rules, actors, and objects (Brembeck *et al.*, 2007). The policy intentions and scientific knowledge of traffic flows and transport economics were key elements in this network. Actors such as the City of Stockholm and the National Road Administration were able to present themselves as credible on the issue, and the trial as a relatively stable project.[6]

The Stockholm trial was put forth as a way to give the inhabitants of Stockholm an opportunity to try the system out; hence, it could also be seen as a demonstration. Demonstrations are usually made in public, in order to enlist spectators and participants as supporters (Hagman and Andréasson, 2006; Henriksson *et al.*, 2011). The interview with Tina was done about a month after the tax had started to be levied. She had already become convinced of its effectiveness. If you see Tina as a spectator, the demonstration was a successful one. Charging worked, and she understood this. But it is also interesting to note the trace of skepticism when Tina says it has become even more attractive for her to drive a car "just because she can afford to pay." She does not seem convinced that this alone is positive and points out that there is a contradiction behind this, in terms of distributional effects.

What was the impact of the charges? Traffic reduction was actually achieved. Traffic counts and surveys showed that 90,000 car trips per day, 20 percent of the traffic in and out from the inner city, disappeared when the congestion charge was introduced.[7] The decrease was due both to modal shifts from car to public transport and to a reduction in leisure trips. People changed their destination for leisure trips or organized their travel in different ways. After the trial, public opinion on the congestion charges shifted from mainly negative to positive, at least among inhabitants of the Stockholm City, who eventually voted "yes" in a referendum on whether to make the charges permanent or not.[8] One probable reason for the successful implementation of the congestion charges is that there was already discontent among the general public with respect to the traffic situation in Stockholm. Thus, the charges were perceived not only as a whip aimed at decreasing the environmental burden from private cars but also as a way to deal with the overall traffic situation. This is supported by the fact that many of those critical of the charges provided detailed ideas for alternative measures.

After the trial, there was a break in congestion charging for about a year. But since the summer of 2007, the congestion tax has been collected continuously, in the same way as it was during the trial. And at the time of writing, in 2012, the number of cars passing the charging cordons is still below the level before the trial. Whether this is due to the charges or whether the explanation is to be sought elsewhere is hard to tell, partly because of the multiplicity of alternative explanations such as the financial crisis, the growing popularity of cycling, road construction projects, etc., but also because we cannot know what the situation would have been today without the tax. However, what we do know is that incentives that remain the same for a longish period of time tend to be internalized, which means that they lose their power as "whips." From that perspective, it is interesting to note that the support for the charges is greater than ever, at least among people living in the City of Stockholm, and the political interest in this type of measure remains high.

7.6 CONCLUDING DISCUSSION

In terms of material space, in our first example (GlashusEtt) the building acts on a discursive level as a representation of sustainability. GlashusEtt does not only say "Hi!" to the people passing by; rather, it reminds them that they are living in what should be a sustainable area: "You should source-separate your waste," whispers GlashusEtt, "because Hammarby Sjöstad is not like any other area; it is special." To its visitors, GlashusEtt focuses less on the practices of inhabitants and more on showcasing the entire area as an excellent example of sustainable urban development: "Look at all these smart solutions!" the building marvels, and then lets people know that they could have these solutions too. Thus, the existence of the building itself has ameliorated the need to use information and marketing or branding to denote Hammarby Sjöstad as a good place. The building is part of the discursive outreach that has led to the good reputation of the area in terms of sustainable urban development.

Our second case was the major urban development area in the making that is the Royal Seaport, where the ideals of urbanity, sustainability, and healthy, livable environments are not tacitly disseminated through an architectural materiality, as in Hammarby Sjöstad, but clearly formulated in text, and published as a vision in a planning document. This has spurred a discursive struggle that recalled the contested Swedish history of social engineering and evoked feelings of surveillance and control. An at-first-sight ordinary newspaper article functioned as the starting point for constructing a counter-narrative that successfully undermined efforts by planners to make people live an environmentally friendly lifestyle.

From the perspective of promoting more sustainable practices, the third case – the congestion charges – is a success story. One explanation for this is the construction of the introduction of the charges as a trial. Thus, it can be argued that testing a reform – through a full-scale demonstration of the reform in question – has the potential to partly hinder or delay the creation of counter-narratives. In the case of congestion charging, automatic cameras, pay boxes, and direct debit concessions were parts of this process. They also involved the creation of new links between the state, administrators, and citizens, and at the same time modified existing links. By creating a socio-technical system of cameras, boxes, administrators, and citizens, the trial also functioned as a way to build momentum for public approval of the congestion charges. When the trial ended, most of the investments needed to make the charges permanent had already been made. In this sense, the trial worked as a creator of path dependency. This case also shows how technological objects and infrastructures have the potential to institutionalize power relations and social relations (Latour, 1998).

Materialities of different kinds have effects. A technical device, once installed, can be accepted, owing to the ease with which it can be forgotten; a critical article can put a policy project on hold; and a building can be a discourse representative as well as a performative space without anybody recognizing it as such. To capture and understand these effects, we argue that any "big" explanations – such as references to society, the market, natural relationships, or the social – must be complemented by the up-close scrutiny emphasized by the actor–network approach.

Policies are instruments of power. In our analysis, the hetero-geneous and situated power of spaces and objects has been advanced, and the spatial power dimensions have emerged as relational and productive. Power is not an object that one can have and hold, but is a relational concept that can be forcing as well as liberating – that is, either "power to" or "power over" (Foucault, 2010). This means that power is not only repressive from a top-down perspective, but also empowering from below through user practices and strategies. An actor–network approach underscores both the possibilities and the difficulties of trying to change practices of everyday life in the city. Those with aspirations to change practices (top-down or bottom-up) are often confronted by numerous obstacles in terms of hindering actors, material conditions, and institutions. One common factor behind these obstacles seems to be the dominating discourse of individual consumer choice. Consumers do indeed have power, but it is a circumscribed power, as the possibilities for going green are highly dependent on the material structures provided by policy and planning and the social context.

To refer only to the individual consumer discourse would, however, be to fall into the *big explanation* trap again. In order to understand how the discourse works, which sociomaterial entanglements it rests upon, and how these could be altered, we need to make use of an approach that closely examines routines, procedures, the use of objects, and spatial practices, as well as spatial configurations both in the present and in relation to past setbacks and successes. Only through acknowledging citizens' inertia to change as being in every sense rational, and then exploring the underpinnings of this rationality, can opportunities for achieving more sustainable everyday life practices be identified.

NOTES

1   Hirdman (1989) has investigated the early history of the Swedish welfare state and argues that it was imbued with an idea of governing life through adjustment, conveyed through normative information campaigns. See also Mattsson and Wallenstein (2010) for connections between architecture and welfare politics, as well as Chapters 2 and 4 of this volume.

2   This type of reformist approach to sustainable development has been heavily criticized by political ecologists (e.g. Dryzek, 2005; Krueger and Gibbs, 2007).

3   According to Axelsson (2012), the environmental footprint of an average Swede is 5.88 gha (global hectares). In contrast to its "green" reputation, the footprint of Stockholm is even higher: on average 6.1 gha. This can be compared to the maximum 1.8 gha per person available to ensure a sustainable and just distribution and use of the Earth's resources.

4   In fact, to demolish a building one needs a permit and it takes effort by humans (demolition workers, engineers, etc.) and non-humans (machines and tools, etc.).

5   The Stockholm trial included congestion charges, new park-and-ride facilities, and enhanced public transport (in particular, new bus routes introduced in autumn 2005). The automatic payment stations formed a charging cordon around the inner city. The area inside – that is, the inner city – was called the charging zone. All vehicles that passed over the cordon during fee time – weekdays from 6:30 a.m. until 6:30 p.m. – were recorded by the cameras. During the evening and night, and at weekends, it cost nothing to drive. Each passage into or out from Stockholm cost 10, 15, or 20 Swedish kronor, depending on the time of day. The maximum amount per day per car was 60 kronor (roughly €7). With only minor adjustments, these arrangements were subsequently made permanent.

6   To go deeper into the political project and its leading players and circumstances, see Gullberg and Isaksson (2009).

7   Information on traffic measurements was released as a monthly publication on the Stockholm trial official website. It is still available at www.stockholmsforsoket.se, through "Evaluation Reports" and then "Monthly Summaries." On the same page, the final report can also be found: *Final Report – The Stockholm Trial, Dec 2006* (Miljöavgiftskansliet/Congestion Charge Secretariat 2006).

8   The inhabitants of the surrounding municipalities did, however, reject the congestion charges in their own self-initiated non-binding referenda on the issue, which were nevertheless not in any way taken into consideration by the national government.

REFERENCES

Axelsson, K. (2012) *Global miljöpåverkan och lokala fotavtryck. Analys av fyra svenska kommuners totala konsumtion* (Global environmental impact and local footprints: An analysis of the total consumption of four Swedish municipalities), SEI Report. Retrieved from www.sei-international.org/publications?pid=2095 (accessed 15 October 2012).

Brembeck, H., Ekström, K. M., and Mörck, M. (2007) "Shopping with humans and non-humans," in H. Brembeck, K. M. Ekström, and M. Mörck (eds.) *Little Monsters: (De)coupling Assemblages of Consumption*, Berlin: LIT.

Bylund, J. R. (2006) "Planning, projects, practice: A human geography of the Stockholm local investment programme in Hammarby Sjöstad," dissertation, Stockholms universitet.

Czarniawska, B. (2005) *En teori om organisering* (A theory of organizing), Lund: Studentlitteratur.

Czarniawska, B. and Hernes, T. (eds.) (2005) *Actor-Network Theory and Organizing*, 1st ed., Malmö: Liber.

Dryzek, J. S. (2005) *The Politics of the Earth: Environmental Discourses*, 2nd ed., Oxford: Oxford University Press.

Foucault, M. (2010) *Säkerhet, territorium, befolkning: Collège de France 1977–1978* (Security, territory, population, Collège de France 1977–1978), Stockholm: Tankekraft.

Green, A. (2006) "Hållbar energianvändning i svensk stadsplanering: från visioner till uppföljning av Hammarby sjöstad och Västra hamnen," 1st ed., dissertation, Linköpings universitet.

Gullberg, A. and Isaksson, K. (2009) "Fabulous success or insidious fiasco: Congestion tax and the Stockholm traffic dilemma," in A. Gullberg and K. Isaksson (eds.) *Congestion Taxes in City Traffic: Lessons Learnt from the Stockholm Trial*, Lund: Nordic Academic Press, pp. 11–204.

Hagman, O. and Andréasson, H. (2006) *Vägvisare mot en hållbar stad? En intervjustudie i tre bostadsområden i samband med Stockholmsförsöket 2005–2006* (Signs toward a sustainable city? An interview study in three housing areas in connection to the Stockholm trial 2005–2006), Gothenburg: Avdelningen för teknik- och vetenskapsstudier, Göteborgs universitet.

Henriksson, G. (2008) "Stockholmarnas resvanor. Mellan trängselskatt och klimatdebatt" (Travel habits of Stockholmers: Congestion charging and climate debate), dissertation, Lunds universitet, 2008.

Henriksson, G., Hagman, O., and Andréasson, H. (2011) "Environmentally reformed travel habits during the 2006 congestion charge trial in Stockholm: A qualitative study," *International Journal of Environmental Research and Public Health*, 8(8): 3202–3215.

Hirdman, Y. (1989) *Att lägga livet tillrätta: Studier i svensk folkhemspolitik* (To adjust life), Stockholm: Carlsson.

JM (2012) "Lev livet i en miljöstadsdel i världsklass" (Live life in a world-class sustainable district). Retrieved from www.jm.se/bostader/sok-bostad/stockholm/stockholm/ hjorthagen/norra-djurgardsstaden-omradesinformation (accessed 28 June 2012).

Johannisson, K. (2010) "Hälsa som påbjuden livsstil" (Health as prescribed lifestyle), *Dagens Nyheter*, 18 November 2010.

Kay, A. (2005) "A critique of the use of path dependency in policy studies," *Public Administration*, 83(3): 553–571.

Krueger, R. and Gibbs, D. (eds.) (2007) *The Sustainable Development Paradox: Urban Political Economy in the United States and Europe*, New York: Guilford Press.

Latour, B. (1998) "Teknik är samhället som gjorts hållbart" (Technology is society made durable), in *Artefaktens återkomst. Ett möte mellan organisationsteori och tingens sociologi* (The return of the artifact: An encounter between organizational theory and the sociology of objects). Stockholm: Nerenius & Santérus.

Latour, B. (2005) *Reassembling the Social: An Introduction to Actor-Network-Theory*, Oxford: Oxford University Press.

Law, J. (2002) "Objects and spaces," *Theory, Culture and Society*, 19(5–6): 91–105.

Mattsson, H. and Wallenstein, S. (ed.) (2010) *Swedish Modernism: Architecture, Consumption and the Welfare State*, London: Black Dog.

Miljöavgiftskansliet/Congestion Charge Secretariat (2006) *Facts and Results from the Stockholm Trial: Final Version: December 2006*, City of Stockholm. Retrieved from www.stockholmsforsoket.se/upload/Sammanfattningar/English/Final%20Report_T he%20Stockholm%20Trial.pdf (accessed 15 October 2012).

Pierson, P. (2000) "Increasing returns, path dependency, and the study of politics," *American Political Science Review*, 94 (2): 251–267.

Stockholms stad (1997) *Miljöprogram för Hammarby Sjöstad* (Environmental program for Hammarby Sjöstad), Stockholm: SBK, Miljöförvaltningen, and GFK.

Stockholms stad (2009) *Norra Djurgårdsstaden, Stockholm Royal Seaport, Vision 2030*, Stockholm.

Stockholms stad (2012) *Norra Djurgårdsstaden, Stockholm Royal Seaport, Hjorthagen: Towards a World-Class Stockholm*, Stockholm.

Svane, Ö., Wangel, J., Engberg, L., and Palm, J. (2011) "Compromise and learning when negotiating sustainabilities: The brownfield development of Hammarby Sjöstad, Stockholm," *International Journal of Sustainable Urban Development*, 3(2): 141–155.

Tottmar, M. (2010) "Press på boende om sund livsstil" (Inhabitant pressure on sound lifestyles), *Dagens Nyheter*, 9 November.

CHAPTER 8

FROM ECO- MODERNIZING TO POLITICAL ECOLOGIZING

FUTURE CHALLENGES FOR THE GREEN CAPITAL

Karin Bradley, Anna Hult, and Göran Cars

8.1 INTRODUCTION

STOCKHOLM IS A CITY with significant green and blue qualities. The city is built on islands and there are plenty of green areas. Stockholm promotes itself as a "green" city, where air and water are clean, building and construction reflect environmental concerns, the recycling system is well developed, and a large share of residents use the public transport system. These achievements are sometimes taken as confirmations of Stockholm's being a *sustainable* city. Without devaluing Stockholm's achievements, we argue that sustainable urban development demands much more than new smart technologies and innovative solutions to urban construction and development. Reading the plans and visions of the City of Stockholm from a political ecology perspective – that is, through posing questions as to the implications of the "greening" of Stockholm for other areas, peoples, and resources – provides insights into some of the future challenges that Stockholm will have to face in order to retain its position at the forefront of sustainable urban development.

Stockholm is indeed proud of its green qualities, and its environmental achievements are used strategically in the marketing of the city. The city's environmental qualities have gained international recognition – for instance, the city was awarded the title of European Green Capital in 2010. Completed and planned urban development projects, well-developed public infrastructure, and the natural setting have all given the city an international reputation of being at the forefront of sustainable urban development.

Stockholm's achievements and reputation provide many advantages. First, in a globalized world characterized by increasing competition between cities, a high profile in environmental issues can give a competitive edge by contributing to the image of the city as attractive to existing, as well as potential, residents. Awareness of environmental problems has increased substantially over recent decades, and public policies to promote environmentally friendly solutions can, besides reducing environmental impacts, also promote the increased attractiveness of the region as a living environment. Second, policies promoting environmentally friendly development have created incentives for various actors to develop methods and products that are eco-profiled. This has led to considerable growth in the export of eco-branded Swedish know-how and products. In 2008, Swedish exports related to the environmental sector were valued at 42 billion Swedish kronor (SEK) (equivalent to approximately €4.75 billion), a growth rate of more than 60 percent since 2002 (Kairos Future, 2010).

The inaugural title of European Green Capital highlighted Stockholm as a role model for environmental standards and provided the city with an opportunity to showcase achievements and share experiences internationally. Yet despite the accolades, in this chapter we argue that work being undertaken toward sustainability in Stockholm needs to be sharpened and needs to address more fundamentally political concerns. Using a political ecology perspective, we ask the following questions: How is Stockholm's success story constructed? What spaces and resources are included and excluded in sustainability calculations? What implications does the "greening" of Stockholm have for other areas? How can more environmentally just development be promoted? With the help of these questions, we explore the challenges that Stockholm faces, and the approaches that can help the City to be considered a forerunner also in the future. In the penultimate section of this chapter, we outline a number of those challenges. However, first we make a brief sketch of the more immediate structural challenges facing Stockholm in 2012, describe Stockholm's current plans and visions, and then clarify our usage of the concepts of ecological modernization and political ecology. In the sixth, and final, section, we summarize the outlined challenges and discuss what they might mean for the way in which Stockholm participates in the global "sustainability trade" and knowledge exchange.

8.2 A GROWING BUT STRAINED STOCKHOLM

Since around 2007, Europe (as well as many other parts of the world) has been severely affected by global financial crises. Greece, Spain, Ireland, and France all have high levels of debt, and governments as well as citizens in those countries struggle to pay their debts and find sources of income and affordable housing. Since 2008, the economic crisis, coupled with widespread awareness of climate change and global environmental challenges, has spurred debates about, and research into, rethinking societal organization, politics, lifestyles, housing provision, urban–rural relations, and (most notably) economies in a way that does not rely on traditional, supposedly everlasting economic growth. Furthermore, socioenvironmental movements such as the Transition movement have grown stronger in recent years, demanding both a more people-centered economy and a rethinking of the current trajectories of development, growth, and resource overuse (Hopkins, 2011).

Sweden – and in particular Stockholm – has, however, nearly managed to bypass the economic crises. Few manifestations of the global economic crises evidenced in cities in Central and Southern Europe, or the energetic social movements accompanying them, are seen in Stockholm. The urban development debate is there instead mainly concerned with how the city should grow, and how the municipality should build more and more rapidly with respect to everything from new housing and highway and public transport extensions, to shopping malls and kindergartens. Stockholm is in an era of growth: growth of citizens, visitors, businesses, and traffic. The City of Stockholm arranges exhibitions with the tagline "Stockholm is growing," and Stockholmers are being taught that a bigger and denser Stockholm is both desirable *and* sustainable.

At the same time, Stockholm is facing a number of structural challenges. Social segregation is increasing and living conditions in many of the large-scale postwar suburbs have worsened in recent years.[1] The development of the public transport system has not been able to keep pace with the growing number of users; citizens are upset by delays affecting commuter trains and by packed state subway cars.[2] A major infrastructural investment is under way which will expand the motorway mileage in the region, namely the *Stockholm Bypass*, creating a heated debate among researchers and citizen groups (Finnveden and Åkerman, 2009). Recent visible investments in public transport have been primarily focused on the inner city; for instance, a modern tramway has been constructed between the city center and the green park of Djurgården, where the city's main tourist attractions are located, making many suburban Stockholmers wonder whether the green profile is mainly an *image* of sustainability geared toward tourists. The City announces "cycling projects" from time to time, and has indeed constructed a number of cycle paths. However, compared to cities like Copenhagen and Amsterdam, Stockholm's cycle routes are irregular, dangerous, and in parts simply non-existent.

The city administration argues that Stockholm has managed to reduce its greenhouse gas emissions per capita. However, more recent figures show the opposite: that Stockholmers, when their consumption is included in the calculation, have rather increased their emissions (see further in what follows). These increased emissions lead to environmental effects that might not be visible locally, but affect other countries and regions. To add to the list of structural challenges, Stockholm is also facing an increasingly strained housing market, where young and newly immigrated residents in particular have difficulty in accessing secure and affordable housing, while other households increasingly find themselves with high levels of debt due to high real estate prices (Andersson *et al.*, 2007). Like many other European cities, Stockholm thus faces major challenges in terms of meeting its ecological responsibilities and achieving social inclusion, economic stability, and well-being.

8.3 FRAMING THE GREEN CAPITAL

The environmental achievements realized in Stockholm in recent decades, coupled with the ambitions expressed in plans for the Royal Seaport, and in major policy programs such as the city's trademark, "Capital of Scandinavia," and its *Vision 2030*, indicate that Stockholm aspires to be at the forefront of the international urban sustainability league. The future challenges that, we argue, Stockholm has to take into consideration should be seen in relation to current plans and visions for Stockholm, and efforts made to position Stockholm (and Sweden) as a forerunner in sustainable urban development. We therefore provide a brief overview of the urban development districts promoted by the City, the plans and vision put forward, and national efforts to promote Swedish know-how internationally.

Best-practice urban districts in Stockholm

The emergence of Stockholm's international reputation in sustainability can be traced back to the decision to build Hammarby Sjöstad. A master plan for the area was presented in 1991 and the first residential buildings were completed in 1994. In 1996, the City Council decided that Hammarby Sjöstad should be constructed as a role model for eco-friendly urban development. This meant that technical installations, buildings, and transport systems were adjusted to meet high environmental demands. Hammarby Sjöstad was successful in meeting this objective. The houses used half the energy and water of an average Swedish property. The so-called Hammarby Model, which shows how sewage processing and energy systems interact and how refuse can be handled effectively, was developed (as described in Chapter 4). The achievements in Hammarby Sjöstad have played an important role in the external marketing of Stockholm as a green city.

As the years have passed, the innovative approaches of Hammarby Sjöstad have gradually lost their luster. New initiatives that were developed in Hammarby Sjöstad have become standard in many projects, and new approaches to planning and construction have advanced sustainable construction beyond the point that it was at when Hammarby Sjöstad was planned. Stockholm Royal Seaport is the successor to Hammarby Sjöstad. Starting in 2010, the development of the area is planned to continue until 2025. The idea is that Stockholm Royal Seaport should benefit from environmental experiences drawn from Hammarby Sjöstad. The ambition, as expressed by the City, is to develop "an environmentally sustainable city district with a genuine city environment [that] puts extra demands on technological innovations, building work using energy efficient materials in building, as well as on finding new ways of handling energy as a whole" (Stockholms stad, 2012b).

Stockholm's plans and visions

Parallel to the stated ambitions to develop a green image for Stockholm, a series of other images of the city are promoted. To the annoyance of many other metropolitan regions in northernmost Europe, Stockholm actively brands itself as "the Capital of Scandinavia." This capital claim is based on the argument that Stockholm, with the largest gross regional product and the greatest number of multinational companies, is Scandinavia's economic center and will be Northern Europe's financial center. Aside from its financial strength, Stockholm is also considered to be Scandinavia's most trendsetting cultural city (Stockholms stad, 2012b).

Besides the trademark "Capital of Scandinavia," the city has also developed a long-term vision, *Vision 2030* (Stockholms stad, 2007). A key ingredient in this strategy is to reinforce ongoing regionalization. The vision glimpses into the future and pictures the Stockholm–Mälar region as a place where international high-tech companies work side by side with small spin-offs in the service sector. It is a future where the creative sector is developing successfully, and culture, sports, and entertainment are creating jobs, export opportunities, and growth. Stockholm's low municipal tax is a further stimulant, as is the efficient network of roads, railways, enlarged airports, and harbors. The future city also offers what may be the world's best IT infrastructure and a reliable energy supply system. Stockholm is, in this vision, Northern Europe's number one financial services city, facilitating the supply of capital for many companies. The vision also pictures the emergence of an environmentally sustainable city where "[i]nnovations have resolved many

Figure 8.1 Excerpt from Stockholm's *Vision 2030* document
Stockholms Stad (2012e)

environmental problems, and the city is well on the way to achieving its goal of being fossil fuel-free by 2050." Further, energy consumption has been reduced significantly, eco-friendly methods and materials are used for all new buildings, and "the region's population increase has had little or no negative effect on the local environment, making Stockholm an international role model" (Stockholms stad, 2012c).

The line of thought expressed in the *Vision* is followed up in the comprehensive plan for Stockholm, *Promenadstaden* (the Walkable City), which was adopted in 2010. In parallel to stressing sustainability, the plan recognizes that concepts such as "sustainable growth and sustainable development are problematic because there are no set definitions and because they contain a number of inherent conflicts" (Stockholms stad, 2010). A number of conflicts are brought up in the plan, such as the conflict between improved access and new infrastructure on the one hand and negative environmental impacts on the other, and the tension between the need for new development and construction vs. the ambition to preserve green space.

National efforts toward international branding

Clearly, Sweden has been able to brand itself as a sustainable forerunner for reasons that have a specific and complex cultural and political history. Substantial governmental efforts have been made in the past decade, often operating through private–public partnerships, to enhance and package such a brand (Hult, 2013). Since 2007, the overarching branding concept has gone under the name of *SymbioCity*, an initiative undertaken by the Swedish government through the Swedish Trade Council to package Swedish urban sustainable development for worldwide marketing. SymbioCity contains working models to identify synergies between both technological systems and different public and private stakeholders. Through SymbioCity, it has become possible to find a new way to relate technology to urban development and hence boost the export of Swedish engineering products and services to countries facing rapid urbanization, such as China (ibid.).

Today, all Swedish embassies provide brochures and PowerPoint presentations that communicate the concept of SymbioCity. In concrete terms, it has become the official national platform to market "environmentally friendly" technological products and services. Swedish political institutions together with private companies and PR bureaus have identified a significant opportunity to combine goodwill with opportunities for profitable trade – linking the export of eco-profiled technology to urban development. There has been a strong political will in the past decade to strengthen Sweden's international position in environmentally friendly technology. This has not only been expressed in the efforts to shape a common platform; new institutional bodies have also been initiated both in Sweden (such as the Delegation for Sustainable Cities in 2008 and MISTRA Urban Futures in 2010) and abroad (such as CENTEC, the Centre for Environmental Friendly Technology in 2008 at the Swedish embassy in Beijing). As a result of these efforts, Sweden has been able to showcase existing urban districts such as Hammarby Sjöstad and the Royal Seaport in a larger framework, positioning Sweden and Stockholm at the global forefront of sustainable urban development.

8.4 FROM LIGHT GREEN SUSTAINABILITY TO POLITICAL ECOLOGY

The idea of "sustainability" has gone mainstream. What began as a grass-roots movement to promote responsible development has now become a bullet point in corporate eco-branding strategies. Sustainable development has in recent decades become the catchword within most policy documents and visions of political parties across the spectra, at least in the Global North. While the term has "won the battle of big public ideas," as Campbell formulates it (1996, p. 312 in Kreuger and Gibbs, 2007), there is no consensus on the exact meaning of the concept; sustainability is rather the axis around which discussion occurs, and limits are nowhere to be seen (Kreuger and Gibbs, 2007; Dryzek, 2005, p. 12).

The meteoric rise of the concept of sustainable development is likely due to the fact that it can be easily hijacked for other purposes. Within the past 25 years, sustainability as a concept had its breakthrough, became problematized in world politics and planning, and soon became overused and abused by all political parties, revealing little about the actual differences between their programs. In *Hijacking Sustainability* (2009), Parr describes how the goals of an environmental movement came to be mediated by corporate interests, government, and the military. Parr argues that sustainability is continuously commodified; the more that mainstream culture embraces the sustainability movement's concern over global warming and poverty, the more "sustainability culture" advances the profit-maximizing values of corporate capitalism.

As sustainability represents "the good," it is difficult to oppose. Thus, sustainability serves the purpose of being a common term in policy-making and vision statements, as everyone can agree upon it. But as the term is no longer especially politically charged and no longer expresses any particular meaning except "good," this also means that its content, while formulated as a "sustainable development vision," to a large degree remains up to individual actors to define.

Today, economic growth is generally considered the main priority of nation-states; the discourse of sustainability therefore to a large extent deals with strategies concerning how to reconcile reoccurring conflicts between environmental and economic goals. According to Dryzek (2005, p. 12) the discourse of sustainability problem solving could to a large extent be defined by imaginative attempts to dissolve the conflict between environmental and economic values.

One school of thought that claims that such a conflict can be resolved is that of *ecological modernization*, which argues that environmental improvement can take place in tandem with economic growth. Within the ecological modernization framework, this problem is solved by explicitly describing environmental improvements as being economically feasible; indeed, entrepreneurial agents and economic/market dynamics are seen as playing leading roles in bringing about needed ecological changes (see Fisher and Freudenberg, 2001; Spaargaren and Mol, 1992). The belief is that there is no contradiction between development, growth, and the exploitation of resources on the one hand, and environmental concerns on the other hand. New technology is often seen as the solution, as well as consumption of goods such as electric cars, water-saving showerheads, and ecological products (Hobson, 2006). From the perspective of ecological modernization, it is assumed that further advancement of technology and industrialization will deliver "sustainable growth."

Political ecology is a school of thought skeptical concerning ideas of ecological modernization. The theoretical field of political ecology concerns how socioeconomic arrangements, politics, and environmental aspects are interrelated (Walker, 2005; Latour, 2004). Focus is placed upon understanding local and global flows as well as on interdependencies and power relations between humans and the non-human world. Keil (2007) calls for an urban political ecology that points to sustainability as an elusive goal as long as fundamental processes of uneven development and unequal exchange are not changed. The questions that need to be asked relate to how to balance the processes of uneven development (ibid.). Analyzing the urban metabolism and asking what counts, who and what are included and excluded, and who wins and who loses as a consequence of metabolic flows is thus central to a political ecology approach. With diffuse links between the local environment and the larger ecological system, it becomes crucial to tie these entities together – seeing how ecological impact and local and international (in)justices are interwoven. As Harvey states,

> [w]e cannot reasonably argue for high environmental quality in the neighborhood while still insisting on living at a level which necessarily implies polluting the air somewhere else. We need to know how space and time get defined by the quite different material processes which give us our daily sustenance.
>
> (1996, p. 233)

What interpretations of sustainability can be seen in Stockholm's strategies and how might a political ecology perspective be useful in progressing toward a more environmentally just development?

8.5 CHALLENGES FOR A MORE ENVIRONMENTALLY JUST DEVELOPMENT

Stockholm's plans and framing in terms of international sustainability contain clear echoes of ecological modernization. Overall, the tone assures us that environmental problems can be solved through innovations and efficient management, within the current socioeconomic order. Continued economic growth, attractiveness, and benign environmental development go hand in hand. The high environmental ambitions have in fact been beneficial to business development and growth, as they have spurred development in and export of the clean-tech industry. So the story goes.

One of the key reasons given for Stockholm's receiving the Green Capital Award was the "fact" that the city has managed to cut carbon dioxide emissions by 25 percent per inhabitant since 1990. But how are these figures calculated? What spaces and resources are included or excluded? What implications does the "greening" of Stockholm have for *other* areas, peoples, and "natures"? How could more environmentally just development on a global scale be promoted in the local planning of Stockholm? These types of questions can be derived from a political ecology perspective.

In this section, we discuss the organization, planning, and monitoring of development in Stockholm with the help of such questions. Many cities around the world aspire to be "green," "zero-emission," or "one-planet." More sophisticated methods for measuring the ecological footprints of cities,[3] for classifying the socioenvironmental performance of city districts,[4] and for raising awareness about the interlinkages between social justice and environmental responsibility are pushing the sustainable city league to go beyond the eco-marketing of waterfront housing. Departing from the questions above, what would Stockholm need to do in order for its green profile to be taken seriously, now and in the future? Examples of the kinds of issues that the city will need to deal with are discussed under the six headings that follow.

From territorial boundaries to a consumption perspective

That Stockholm has managed to cut carbon dioxide emission by 25 percent per inhabitant since 1990 is indeed admirable. But another picture emerges when a consumption perspective is introduced and questions of what spaces are included or excluded in the calculations are addressed. These aspects are important to take into consideration if the efforts behind Stockholm's environmental targets are to be taken seriously in the future.

An economy that is able to sustain growth in GDP without simultaneously degrading environmental conditions is said to be *decoupled*. This is exactly the discourse that the City of Stockholm taps into when it nominates "sustainable growth" as the overall aim of the comprehensive plan. However, whether or not this is achieved can be questioned. While marketing efforts and governmental initiatives claim emission reductions in Sweden and Stockholm, the decrease in emissions shown in these calculations is also connected to simply moving production out of the country and neglecting imports in the statistics. If instead the calculation is made from a consumption perspective, emissions from all stages of production *from cradle to grave* are allocated to the final consumers of goods and services. Emissions attributable to exports are therefore deducted from Sweden's emissions balance, and emissions generated by imports to Sweden are added in order to estimate emissions from Swedish consumption. The Swedish Environmental Protection Agency has recently shown that, from a consumption perspective, the emissions of Swedes have not decreased but in fact *increased* by 9 percent during the period 2000–2008 (Naturvårdsverket, 2012a). Official measurements from the City of Stockholm account for neither the emissions and resource use generated outside of the city boundaries from transportation into and out of the city (e.g. flights and boat journeys), nor (notably) emissions resulting from the consumption of food and goods produced outside the city (Saar, 2010). Much of the decrease in emissions reported by supposedly decoupled countries can be accounted for in the displacement of environmental impacts to other countries. The usual production-based measurements make a city like Stockholm, which has a predominantly service-based economy, perform well, while if measurements are shifted to take into account the global effects of consumption, Stockholm instead generates a considerably larger carbon footprint, since its residents are prominent consumers. Compared to other large Swedish cities, Stockholmers in fact have the largest carbon footprint, because of larger emissions from food, consumer goods, furniture, and renovation compared to the other cities (SEI, 2012, pp. 8–11).

In the current Environmental Program of the City of Stockholm (2013–2015), several specified goals concern the City administration's own consumption of goods and services. These goals address issues such as increasing the share of eco-labeled products in municipal kindergartens, and lowering waste levels in municipal offices and centers. However, the 59-page document does not contain concrete goals regarding residents' consumption, or statements of what *needs* be done; there are only statements of what supposedly *can* be done – for instance, "Through information campaigns and the City's own work the City *can* inspire and illustrate the goal of preventing waste" (Stockholms stad, 2012, p. 27, emphasis added). This illustrates a belief in voluntary change, placing the responsibility on individuals to act and consume "sustainably" (as described in Chapter 7). One can also note how the City of Stockholm, in its reoccurring survey of citizens' environmental attitudes, did not (among the 51 questions) bother to include questions on the level of consumption, the number or length of flight trips, or the size of housing, but instead concentrated on detailed questions about recycling, eco-labeled products, and time spent in "nature" (Bradley, 2009, pp. 255–256). This contributes to an image of the eco-friendly citizen as a

Table 8.1 Comparison of the national reporting of CO_2 emissions per capita calculated from a production perspective compared with a consumption perspective

Emission of greenhouse gases (tons/year)	Sweden	Gothenburg	Linköping	Malmö	Stockholm
Official national reporting, total (2005)	66,630,190	3,397,100	763,714	1,196,025	2,241,471
Official national reporting, per capita (2005)	7.4	7.0	5.6	4.4	2.9
Emissions per person including consumption (2004)	14.2	13.8	13.5	13.4	15.7
Increase in percentage when consumption is included	93%	98%	143%	204%	439%

Data from SEI (2012)

person who recycles and consumes eco-labeled products, rather than a person who economizes with resources such as flying, the size of housing, or levels of material consumption (ibid.).

Recent figures from the Swedish Environmental Protection Agency (Naturvårdsverket) (Naturvårdsverket, 2008, 2012a) show that when consumption is included in the calculations, emissions in Sweden are equivalent to just over 10 tons of CO_2 per capita. In order to meet the political goals that have been set in Sweden, emissions for an average Swedish consumer need to decrease to half that level by 2020 and to a fifth of that level by 2050 (2008, p. 13).

These types of findings, along with new measuring methods and governmental reports, suggest that Stockholm and other cities with high environmental ambitions must consider relational effects beyond their own municipal boundaries, and include a consumption perspective in goal setting and monitoring, as well as in practical planning and policy making.

While awaiting the introduction of green taxation or other national or international policies, a city and a region can achieve a lot through their own initiatives. For instance, campaigns that do include the issue of consumption could be expanded (Stockholms stad, 2012d). Waste collection fees could be increased. One could use tools such as the notion of *climate debt* (tools that shed light on the responsibilities of different nations to reduce their greenhouse gas emissions) and per capita *ecological footprints* to introduce a consumption perspective. The Stockholm Environment Institute has developed an index specifically for municipalities, which addresses environmental effects from a consumption perspective rather than only considering production (the more common measurement) (Paul *et al.*, 2010).[5] Handling the challenge of consumption and its socioenvironmental impacts is often considered to be beyond the reach of planning and policy. At the same time, the expansion of commercial space in urban planning practice often promotes *increased* consumption and socializing through consumption. Developing and caring for non-commercial spaces and meeting places is therefore important. Today, peer-to-peer sharing, or "collaborative consumption," is a growing phenomenon (see Botsman and Rogers, 2011), and cities can promote collaborative consumption and sharing schemes, not only the usual car and bike sharing schemes but also sharing schemes for goods, tools, and spaces. These are examples of approaches that planners could attend to and make space for. The attempt to influence matters that are seen as "private," such as the consumption of goods, is controversial; nevertheless, such efforts can be compared to the (now acceptable) ways in which planning and policy making attempt to influence private transport in favor of collective and sustainable transport. Why not, therefore, move from supporting private consumption to supporting more collective and sustainable consumption of goods?

From curbing carbon to dealing with "peak everything"

Today, the main focus of the environmental debate is on climate change and the reduction of greenhouse gas emissions. As illustrated in the preceding subsection, environmental performance is generally measured in terms of CO_2, and this is reflected in national as well as local goals and monitoring: Sweden being carbon neutral by 2050,[6] Stockholm being fossil fuel-free by 2050, and Stockholmers having reduced their emissions to less than 3 tons of CO_2 per capita per year by 2015 (Stockholms stad, 2012, p. 18).

Within environmental science, it is, however, well established that carbon emissions and the peaking of fossil fuels are not the only environmental challenges we face. Researchers such as Richard Heinberg (2007) use the term "peak everything" to remind us that the supply of other natural resources will also decline in the twenty-first century – including phosphorus, rare earth metals (e.g. the neodymium used in wind turbines and hybrid cars), silicon (used in solar cells), and copper and cobalt (used in electronics, etc.). This means that much so-called green technology is paradoxically also reliant on the use of scarce, non-renewable natural resources. We are currently witnessing struggles over the control of oil, as well as rising prices; thus, there are clear reasons to move away from oil dependency. Human ecologist Hornborg (2011) argues that in the future we will see ever more competition for other scarce resources that our "green-tech lifestyles" are becoming dependent on. This has already become evident for some "sustainable technologies" – for instance, in the way that the use of ethanol as a "green" fuel in the Swedish car fleet is contributing to environmental degradation, loss of biodiversity, and conflicts in Brazil and other countries, as soy and sugar monocultures expand in order for it to be possible to export biofuels (Galli, 2011).

From this perspective, a forward-looking green city needs to be able to live and prosper with *less* use of scarce and distant resources and without becoming reliant on the use of what today may be considered "green technology." In its goal setting, practical work, and monitoring, such a city needs to go beyond a narrow carbon focus and include other resources as well – and hence promote buildings, technology, and lifestyles that are less reliant on such resources.

From GDP to sustainable prosperity

The common Western notion of linear development – often measured as GDP growth or increased material wealth – has been criticized for decades as lacking a focus on well-being and human development, and for the (over)use of natural resources, as well as for engendering dependency relations between different nations and regions (Skovdahl, 2010; Goossens *et al.*, 2007). Alternative indicators have been developed, such as the Human Development Index[7] (used by the United Nations Development Programme) and the Life Satisfaction Indexes. There are also indicators that combine quality of life with ecological footprints, such as the Happy Planet Index developed by the New Economics Foundation (2009).[8] In recent years, international debate has intensified regarding the necessity of using the latter types of indexes that account for the use of natural resources and well-being – and hence measure, evaluate, and steer politics directly toward "sustainable prosperity," rather than indirectly via GDP (Starke *et al.*, 2012; Naturvårdsverket, 2012b). In Europe, notably in the United Kingdom and France, as well as in the East, in China and Bhutan, governments have for several years been developing such "happiness" indicators (Stiglitz *et al.*, 2009; Naturvårdsverket, 2012b).

In the Happy Planet Index, 151 countries are ranked: Costa Rica gets the best score, Sweden gets an average score (53rd place), the United States is in 105th place, and Botswana in 151st. Overall, Latin American nations show the highest levels of well-being while simultaneously not overusing natural resources – hence juxtaposing the GDP image of what is often considered as "success" with alternative calculation methods.

Developing and applying these indexes of "sustainable prosperity" is also important at the regional and local levels. For a city like Stockholm, there are possibilities to improve the ecological footprint as well as quality of life. In the public debate, it is often assumed that emissions reductions and less resource use will imply "sacrifices" to well-being. For poorer regions, GDP growth (when distributed across the population) has indeed often implied a higher quality of life (coupled with increased emission levels). However, for regions in the global North, higher national income per person over a certain level raises emission levels but is not accompanied by the desired increase in well-being – rather the opposite (Wilkinson and Pickett, 2010, pp. 8–10). In a recent report for the Swedish Environmental Protection Agency, a research team has shown that reduced emissions and resource use in Northern countries have the potential to go hand in hand with increased quality of life (Naturvårdsverket, 2012b). Its report points to three main fields: (1) shorter working hours; (2) urban development that supports less energy use, and public transport; and (3) increased consumption of services. When one relates environmental impacts and the level of well-being experienced to different types of activities, there is a clear positive correlation between activities with low environmental impact (such as spending time outdoors, socializing, or resting) and high levels of well-being. On the other side of the spectrum, car commuting was found to be one of the activities contributing to the lowest level of well-being (ibid., pp. 40–41). Finding local ways to simultaneously combine reduced resource use with increased quality of life would be a way to secure an international reputation as a forerunner.

A recent research project at the Royal Institute of Technology (KTH) supports a critical stance regarding whether or not GDP really captures

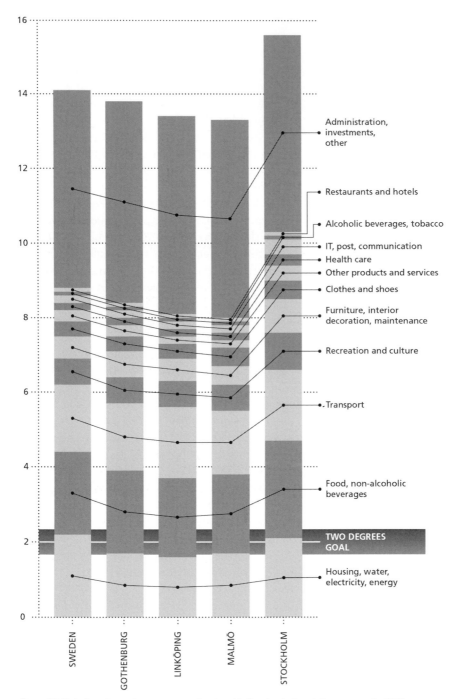

Figure 8.2 Emission of greenhouse gases – direct and indirect emissions in tons per capita, 2004. When indirect emissions from consumption are included, Stockholm shows higher emission levels per capita than other major Swedish cities.

Adapted from Naturvårdsverket (2012b)

prosperity and quality of life. In a longitudinal study, people moving to the Swedish city of Ystad were asked about the motives that guided their decision to move there. The fact that the economic and demographic composition of movers may change over time was taken into account in the project, and found to be of minor importance. The longitudinal figures indicate a significant change in individuals' values. For those who moved to Ystad prior to 1970, 65 percent stated that employment opportunities were the main reason for their locational choice. Between 1970 and 1989, this figure declined to less than 40 percent. For those moving to Ystad after 1990, only slightly more than 20 percent stated that employment opportunities were the major relocation factor. Parallel to this significant decrease in what can be regarded as a GDP-related indicator, other non-economic conditions grew in importance for people when making their locational decision. For those moving to Ystad prior to 1970, 35 percent nominated "urban attractiveness and qualities in the built environment" as the major reason for their move. This figure has gradually increased, and more than 65 percent of those moving to Ystad after 1990 reported urban qualities to be the major reason for their decision to move (Cars, 2006).

This points toward the importance of focusing planning and policy making in a city like Stockholm on "sustainable prosperity" – increased quality of life coupled with reduced ecological footprints – rather than taking the detour via increased gross regional product and simply hoping that this will trickle down to increased quality of life and environmental improvements via a causal link that does not exist anymore, at least not in the Global North (Wilkinson and Pickett, 2010, pp. 8–10).

From globalized service society to localized economies

Today's city planning can be said to respond to the needs of "the service/ knowledge society," doing away with the spatial form of the industrial society, functional separation, and large-scale modernist housing. With inspiration from gurus such as Richard Florida, Stockholm, as well as many other cities, aspires to develop high-amenity, mixed-use urban environments to attract the "creative class," a group employed in knowledge-intensive industries (ICT, biotechnology, banking, and finance), good consumers who keep local service providers busy and the global economy spinning.

However, economies and societal forms never last forever; the agrarian society, the mercantile society, and the industrial society have all had their peaks and declines, and so will the "knowledge/service/consumerist society." While it is often described as eco-friendly, the late-capitalist service society rests on the notion of eternally continuous economic growth, a global division of labor, and production and transport in a global carousel reliant on scarce resources. From a political ecology perspective, it is important to highlight the power relations in these global interdependencies: who are the

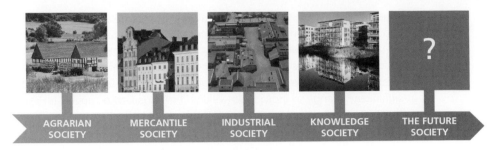

Figure 8.3 Rough sketch of societal organization and spatial form

winners and losers in the global exchange, and what socioenvironmental costs are incurred in the process? As has been pointed out by researchers and environmental organizations, it is generally the low-income nations and societal groups that tend to use the least amount of resources but are most vulnerable to environmental degradation and climate change (Agyeman *et al.*, 2003). The contemporary globalization and growth paradigm is increasingly being challenged by researchers, social movements, and even government advisers (Jackson, 2009; Rockström and Wijkman, 2011; Heinberg and Lerch, 2010; Hopkins, 2011). However, in Stockholm's comprehensive plan *Promenadstaden* (Stockholms stad, 2010) and *Vision 2030*, the envisaged city is clearly a growth-oriented service/knowledge economy. For instance, it is stated that "[t]he city should prepare the way for new offices in locations that are able to attract the high-skilled service sector" – that is, in primarily inner-city locations (Stockholms stad, 2010, p. 16). There are no traces of visions fostering the development of other economies, organizations, or forms of production and consumption. However, changes globally indicate that it is perhaps time for Stockholm to think not only about what spatial forms match the needs of the 24–7 consumerist "knowledge economy" but also of what societal forms might evolve in parallel and become dominant in the future.

In the future, it is likely that prices of energy and hence transport will rise. Under such a scenario, consumer goods currently produced in low-cost countries and transported long distances will become more expensive. At the same time, economic crises and high unemployment rates in Europe might motivate increased regional production of basic goods, such as food, clothes, and utilities. Conscious consumers and socioenvironmental movements are also working toward more locally produced goods and the strengthening of local economies and jobs (Heinberg and Lerch, 2010; Darley *et al.*, 2005; Hopkins, 2011; Murphy, 2008). Fast-growing clothing brands such as American Apparel and Engineered Garments have deliberately located their production in "sweatshop-free" factories in downtown Los Angeles and Brooklyn respectively.[9] These locally produced goods are still, however, currently only niche brands with a limited market. The trend magazine *Monocle* also notes that there is a growing demand not only for restaurants that have open "see-in kitchens," assuring high-quality production and working conditions, but also for firms that pursue "open" production of clothes and other consumer products.

The urban development currently promoted in Stockholm can be characterized as "mixed use." In practice, this means a neat mix of small offices, cafés, shops, services, and housing – but not spaces for production of goods such as clothes, food, and other utilities. Rather the opposite: the areas on the outskirts of the city center that in the past housed maintenance and repair shops, small-scale industries, breweries, etc. are being cleared in order to make space for waterfront housing sprinkled with offices, shops, and cafés. But in the future, for a city like Stockholm to house a sustainable economy it also needs spaces for local production. The Stockholm Chamber of Commerce has noted that demands from companies wanting to move their production from abroad to the Stockholm region already exist, for instance in small-scale textile production.[10] However, these companies currently have difficulties in finding well-located spaces in the Stockholm region, and thus might opt to locate in other cities. In the future, Stockholm may need to develop strategies for creating attractive conditions not only for the ICT, banking, and service sectors but also for the production of basic goods.

From green surfaces to productive landscapes

Stockholm is sometimes called the green and blue city; located on islands, with open spaces accounting for approximately 40 percent of its surface, it is certainly green and blue. In the City's comprehensive plan, greenery is described as spaces for recreation, biodiversity, and leisure. In order to meet many of the future challenges of Stockholm, it is important to consider greenery as also constituting possible productive landscapes, and as possessing benefits additional to those related to biodiversity and leisure.

Industrialization has created a great dependence on fossil fuels in the food chain for the sake of large-scale, cost-efficient processes (Cordell et al., 2009). At the same time, traditional linear consumption flows ("take–make–dispose") are coming up against constraints in terms of the availability of resources. The challenges on the resource side are compounded by rising demands from the world's growing and increasingly affluent population. As a result, one can observe overuse of resources and more volatility in many markets (Ellen MacArthur Foundation, 2011). One such market is food. In recent years, we are, however, also seeing a counter-trend toward the development of urban farming and local food production.

Urban agriculture is certainly not new. Stockholm's first allotment garden was established in 1904. Yet, with the climate crisis and economic crisis as a backdrop and internet access providing the possibility for connection, there is now also a trend toward local initiatives in different parts of the world inspiring each other. In Stockholm, for example, a group of citizens formed a voluntary organization in 2012 to initiate an urban garden on unused rail tracks in the inner city (Eriksdalsspåren), inspired by initiatives such as the movable Prinzessinnengärten in Berlin[11] and the architectural renewal project The High Line in New York.[12]

Detractors of urban agriculture claim that it is idealistic and impractical, providing only a few dinners to an interested, green-washed middle class. Certainly, urban farming is not the answer to all the problems related to the unsustainable overuse of resources. Yet it does provide many diverse benefits, and several small-scale initiatives might bring forth larger shifts in mindset. Besides the more general benefits that urban agriculture can deliver, such as food security (Kortright and Wakefield, 2011), access to nutritious and healthy food (Karanja *et al.*, 2010), personal well-being (Armstrong, 2000), education (van Leeuwen *et al.*, 2010), and the reduction of food miles (Knight and Riggs, 2010), urban farming can contribute to shaping various forms of socioeconomic relations.

The simple fact that people engage in a joint agricultural effort in the city can increase social interaction, resulting in more active communities and stimulating the integration of diverse societal groups (Armstrong, 2000). It can also pay off because of the increased number of *eyes on the street*, a positive influence on neighborhood safety (Armstrong, 2000).

Urban farming can also have direct impacts on urban economies. Urban agriculture is a way to increase the use of space, since it is a land use that can be applied in several ways and scales, filling up voids and otherwise unused plots of land and buildings (Grewal and Grewal, 2011). Additionally, the involvement in and produce from urban agriculture projects can create employment and income opportunities (van Veenhuizen and Danso, 2007). Furthermore, it can lead to a diversification of associated industries, triggering innovations and the emergence of new industries and products in new sectors (Merson *et al.*, 2010). At the same time, local food production can clearly bring about developments that have a less negative impact on the world's environment, ecological systems, and waste cycles. Besides the potential to enhance the city's biodiversity (Faeth *et al.*, 2011), urban agriculture can foster urban greening, which in turn can have advantages for air quality, can reduce urban heat islands and noise (Lin *et al.*, 2011), and can contribute to the recycling of (food) waste, creating shorter cycles of waste processing and reducing the amount of residual waste (Eriksen-Hamel and Danso, 2010).

There are many ways to work toward more local food production. Policymakers and planners have an important role to play in ensuring that land-use plans and regulations enable urban farming activities. Social workers, activists, and citizens are important for initiating and pushing for "unused" sites to be developed into community gardens. By contrast, urban planners, architects, and engineers can contribute to the development of infrastructure to accommodate food production in order to feed the urban users of the future.

The City of Stockholm has the possibility to encourage the growing interest in urban farming and realize its possible multiple benefits in both socioeconomic and environmental terms. The current comprehensive plan does not contain a word about local food production. A possibility therefore clearly exists to develop strategies that view food production as a tool in urban development, not with the main aim of feeding the whole urban area but rather looking to contribute, as a small step, to social inclusion, recreation, well-being, local ecosystem loops, localized production, and the reduction of transport.

From new production to retrofitting

As has already been described, Stockholm and other cities with ambitions to be environmental forerunners often promote themselves by showing examples of new construction branded as "sustainable city districts" or "CO_2-neutral." A focus on novel approaches, methods, and/or technologies for new construction and physical developments is typical. For instance, the focus is often on buildings and their interplay with adjacent surroundings (e.g. through access to green space, services, and transport). New "sustainable city districts" are possibly easier to market, sell, and showcase, compared to the refurbishment of existing areas where one must confront the histories, opinions, and habits of residents. There is, however, one recent exception from this dominating focus on new construction in Stockholm. The city has recently, with support from the national Delegation for Sustainable Cities, launched the project Sustainable Järva (*Hållbara Järva*). The project is targeted at the refurbishment of Järva, a group of large-scale housing estates from the 1960s and 1970s located in the northern part of Stockholm. Besides exhibiting ambitions for improving social conditions, the project also has high environmental aspirations. Järva is to become one of Stockholm's "profile areas" in terms of environmental achievements. Crucial ingredients to realize this ambition are new energy-effective technologies and new concepts for transportation. Ultimately, the goal is for Järva to become a national and international role model for the sustainable renovation of large-scale modernist housing estates from the postwar era (Stockholms stad, 2012).

Apart from this example, the focus of the City's environmental initiatives is overwhelmingly directed toward the construction of *new* buildings. The significance of addressing retrofitting becomes increasingly apparent when considering the volume of new production. The City of Stockholm's housing stock comprises slightly more than 440,000 units. In 2010, 4,090 new units were added, corresponding to slightly less than 1 percent of the existing housing stock (Stockholms stad, 2011). This low figure points to the need to develop comprehensive strategies for existing building stock, as moves toward a more sustainable profile and better environmental performance through new construction *on the margin* are even at best very slow.

8.6 CONCLUDING DISCUSSION: BEING AT THE FOREFRONT INTERNATIONALLY

Sweden and Scandinavia have been widely praised for their efforts to develop and promote models of sustainability for the rest of the world to follow (Kreuger and Gibbs, 2007). International architectural firms headquartered in Sweden are often invited to contribute (via "best practice") to eco-city developments worldwide. As has been shown by Hult (2013), within these projects, Swedish architecture and urban planning firms are driven by the advantage of being able to brand their projects as "sustainable and Scandinavian." Discussions of the value of demonstration projects can be traced to international interest in housing exhibitions such as Bo01 in Malmö and Hammarby Sjöstad in Stockholm. These projects have, in the past decade, become pillars in the success of the Swedish Sustainable City model. Hence, "the sustainable city" becomes a Swedish export product. But what is the basis for this model or brand? And what is needed for Stockholm to continue to be regarded as an international role model in the near and distant future? The SymbioCity concept uses an image that illustrates what is called "the Swedish Experience." This is a graph of increased GDP growth and decreased CO_2 emissions, and it is accompanied by the statement that "[s]ince 1990, CO_2 emissions have been reduced by 9% while the economy has been growing at stable speed."[13]

The idea of increased GDP and decreased CO_2 emissions not only is presented through SymbioCity, but also is a core element in the national marketing material for Sweden (*Sverigebilden 2.0*) and has even been nominated as the basis for the shaping of a new green economy for Sweden.[14] The power of the SymbioCity graph lies in the "fact" that Sweden has managed to achieve a decoupled economy. However, exactly how, whether, or to what extent this has been, or can be, achieved is, as we have shown, a subject of debate, depending on what factors are included. So far, carbon emissions have been calculated from the production side, with the territorial boundaries of the nation-state or the municipality used as the boundaries of calculation. From such a perspective, Sweden – including Stockholm – has experienced a trend toward falling CO_2 emissions since the 1970s. However, in this chapter we have pointed to recent research reports stressing the importance of analyzing greenhouse gas emissions from the perspective of consumption. A consumption perspective provides a more accurate picture of how our lifestyles affect the climate. From such a

SymbioCity SUSTAINABILITY BY SWEDEN

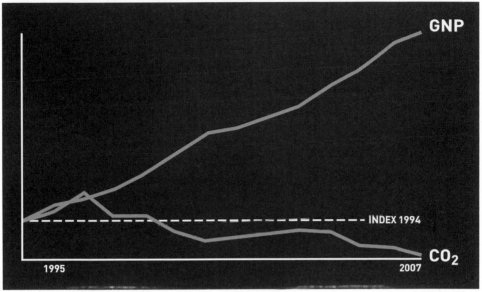

Figure 8.4 The SymbioCity promotion graph "The Swedish Experience," illustrating the phenomenon of a "decoupled" economy

perspective, the emissions of Swedes, and of Stockholmers, have instead *increased* in the past decade (Naturvårdsverket, 2012a). The City of Stockholm has to take both perspectives into consideration. The discussion of who is accountable for impacts elsewhere is crucial in the sustainability debate, and is closely related to the issue of responsibility. In addressing forms of urban development, it is important to consider the consumption patterns and lifestyles of residents, and the way in which urban form, infrastructural systems, and access to commercial and non-commercial spaces influence behavioral patterns. It is, however, important not to confuse the consumption perspective with transferring a major part of the responsibility for CO_2 emissions onto the individual. The City and private development firms have a responsibility to create infrastructure and a built environment that make a sustainable choice the easiest option.

Stockholm's current plans contain clear echoes of ecological modernization. Overall, the tone – both in the overall branding concept of SymbioCity and in the comprehensive plan for Stockholm – assures its audience that the environmental problems can be solved through innovations and efficient management from within the current socioeconomic order. In order to be at the forefront of sustainable development, the City of Stockholm also needs to go beyond the narrow carbon focus and should, in its goal setting and monitoring and practical work, include other resources as well. As such, the City should promote buildings, technology, and lifestyles that are less reliant on such resources. At the same time, we have also noted the importance of moving from GDP to other measures of "sustainable prosperity." Constantly challenging dominant images of development and being open to reformulate future paths is crucial if the City of Stockholm wants to be at the forefront internationally.

Sweden is a progressive country in many ways. In Stockholm, there are 700,000 people using public transport on a daily basis, and during the busiest morning rush hour 78 percent of the trips toward the center of Stockholm are by public transport.[15] There are good traditions in construction and energy-efficient housing worth maintaining and developing further. Technological innovation in heat pumps, solar panels, and waste vacuums can save energy. In urban planning projects, it is often seen as apolitical or uncontroversial to propose technological transfers between nations. Stockholm has an interest in "marketing" the sustainable city as an export product with an emphasis on the decoupling effect in order to increase profits to export-oriented businesses. It feeds on and feeds into an emerging transnational urban sustainability language – a language that is seen as apolitical. Winner, however, makes an interesting observation when noting that

> [p]eople are often willing to make drastic changes in the way they live to accommodate technological innovation while at the same time resisting similar kinds of changes justified on political grounds. If for no other reason than that, it is important for us to achieve a clearer view of these matters than has been our habit so far.
>
> (1980, p. 11)

When working with sustainable urban development projects, it becomes crucial to ask what societal and habitual changes the introduction of a specific technology might bring in different parts of the world. Today, the urban districts of Hammarby Sjöstad and Bo01 in Sweden serve as packaged examples that are ready for export. They are part of a best-practice market, where eco-cities are viewed as urban projects with well-defined borders (in both space and time). As was noted, newly built housing units in Stockholm correspond to slightly more than 1 percent of the existing housing stock; it is therefore crucial to also address retrofitting in terms of sustainable urban development. The question of whom these areas are built for is also of crucial importance. To showcase areas such as Hammarby Sjöstad that are inhabited by a largely well-off and homogeneous population is, from a political ecology perspective, far from sustainable. It is time to move on from project-based showcase urban development districts to continuous processes of urban development.

Compared to many other countries, Sweden has developed processes to include public and private actors in early phases of urban development projects. Sweden also possesses competence in coordinating technical systems such as combined waste and heat systems, as well as in using biogas as fuel for buses. International delegations visiting Stockholm are surely interested in technical solutions, but there is also an increased interest in other issues of urban development – like the processes of coordinating technical systems and the planning process. This is probably where Stockholm has the best chance to serve as an important ground for international exchange toward environmentally just futures, even though the question of implementation is always more politically charged than that of buying technical products. In order to contribute to a more environmentally just development globally from within Stockholm, those working with urban development projects can bear in mind the effect that the "greening" of the city has on other areas, peoples, and resources.

NOTES

1   Even the City Plan of Stockholm (Stockholms stad, 2010) acknowledges the "huge disparities between districts and between the inner and outer city" (p. 10) and states that "[i]nitiatives undertaken to improve the conditions in marginalized areas have not yet been able to reverse these trends" (Summary version, p. 4).
2   See the website of the Stockholm County Administrative Board, www.tmr.sll. se/rufs2010/rufs/Strategierna/Oka-uthallig-kapacitet-och-kvalitet-inom-utbildningen-transporterna-och-bostadssektorn/Oka-uthallig-kapacitet-och-kvalitet-inom-utbildningen-transporterna-och-bostadssektorn/Alltmer-belastat-transportsystem/ (retrieved 1 July 2012).
3   The Stockholm Environment Institute has, for instance, developed the index REAP (Resources and Energy Analysis Programme), which has a lifecycle focus showing the impacts globally of people's consumption, housing, and transport habits. See www.sei-international.org/mediamanager/documents/Publications/Climate/reap.pdf (retrieved 7 April 2012).
4   In the United Kingdom, systems for undertaking sustainability assessments of city districts have been developed (BREEAM Communities and LEED Neighborhood). Currently, work is ongoing to develop a classification system ("Hållbarhetscertifiering av stadsdelar") that is suitable for a Swedish context. See the website of IVL, the Swedish Environmental Research Institute: www.ivl.se/press/nyheter/nyheter/hallbarhetscertifieringavstadsdelarpaframmarsch.5.4c8025261319380cae18000291.html (retrieved 13 April 2012).
5   The index is called REAP (Resources and Energy Analysis Programme). See note 3.
6   This goal was announced in 2009 by the Minister of Environment. See www.energyefficiencynews.com/articles/i/1951/ (retrieved 13 April 2012).
7   The HDI uses three parameters: life expectancy at birth, education level, and standard of living. See http://hdr.undp.org/en/statistics/hdi/ (retrieved 29 June 2012).
8   For more information and rankings of the Happy Planet Index, see www.happy planetindex.org/about/ (retrieved 29 June 2012).
9   On the website of American Apparel, there is detailed information about its Los Angeles factory and working conditions: www.americanapparel.net/factorytour.html (retrieved 13 April 2012).
10   Anna Wersäll, Stockholms Handelskammare, presentation at Stadsbyggnadsdagen, Upplands Väsby, 25 August 2010.

11 See www.prinzessinnengarten.de (retrieved 4 June 2012).
12 See www.thehighline.org (retrieved 4 June 2012).
13 See www.symbiocity.org
14 See www.framtidskommissionen.se/fyra-utmaningar-fyra-utredningar/gron-tillvaxt (retrieved 20 June 2012).
15 The website of the City of Stockholm, http://international.stockholm.se/Politics-and-organisation/A-sustainable-city/ (retrieved 1 July 2012).

REFERENCES

Agyeman, J., Bullard, R. D., and Evans, B. (2003) *Just Sustainabilities: Development in an Unequal World*, London: Earthscan.
Andersson, Å. E., Pettersson, L., and Strömquist, U. (eds.) (2007) *European Metropolitan Housing Markets*, Berlin: Springer.
Armstrong, D. (2000) "A survey of community gardens in upstate New York: Implications for health promotion and community development," *Health and Place*, 6: 319–327.
Botsman, R. and Rogers, R. (2011) *What's Mine Is Yours: How Collaborative Consumption Is Changing the Way We Live*, London: Collins.
Bradley, K. (2009) "Just environments: Politicising sustainable urban development," doctoral dissertation, KTH (Royal Institute of Technology), Stockholm.
Brundtland Report (1987) *Our Common Future*, World Commission on Environment and Development, Oxford: Oxford University Press.
Cars, G. (2006) *Kultur, turism och stadsattraktivitet. Kultur som attraktion och värdeskapare*. Stockholm: KTH.
Cordell, D., Drangert, J.-O., and White, S. (2009) "The story of phosphorus: Global food security and food for thought," *Journal of Global Environmental Change*, 19(2): 292–305.
Darley, J., Room, D., and Rich, C. (2004) *Relocalize Now! Getting Ready for Climate Change and the End of Cheap Oil*, Gabriola Island, BC: New Society.
Dryzek, J. (2005) *The Politics of the Earth: Environmental Discourses*, 2nd ed., Oxford: Oxford University Press.
Ellen MacArthur Foundation (2011) *Towards the Circular Economy*, no. 1: *Economic and Business Rationale for an Accelerated Transition*, Cowes, UK: Ellen MacArthur Foundation.
Eriksen-Hamel, N. and Danso, G. (2010) "Agronomic considerations for urban agriculture in southern cities," *International Journal of Agricultural Sustainability*, 8(1–2): 86–93.
Faeth, S. H., Bang, C., and Saari, S. (2011) "Urban biodiversity: Patterns and mechanisms," *Annals of the New York Academy of Sciences*, 1223: 69–81.
Finnveden, G. and Åkerman, J. (2009) *Förbifart Stockholm, miljön och klimatet. En fallstudie inom vägplaneringen*. Stockholm: KTH. Online, available at: www.infra.kth.se/fms/pdf/Forbifart_Stockholm_miljon_och_klimatet.pdf (accessed 1 December 2012).
Fisher, D. R. and Freudenburg, W. R. (2001) "Ecological modernization and its critics: Assessing the past and looking toward the future," *Society and Natural Resources*, 14: 701–709.
Galli, E. (2011) "Frame analysis in environmental conflicts: The case of ethanol production in Brazil," dissertation, KTH, Stockholm.
Goossens, Y, Mäkipää, A., Schepelmann, P., and van de Sand, I. (2007) *Alternative Progress Indicators to Gross Domestic Product (GDP) as a Means towards Sustainable Development*, study provided for the European Parliament's Committee on the Environment, Public Health and Food Safety, Brussels: Policy Department A: Economic and Scientific Policy.
Grewal, S. S. and Grewal, P. S. (2011) "Can cities become self-reliant in food?" *Cities*, 29: 1–11.
Harvey, D. (1996) *Justice, Nature and the Geography of Difference*, Malden, MA: Blackwell.
Heinberg, R. (2007) *Peak Everything: Waking Up to the Century of Declines*, Gabriola Island, BC: New Society.

Heinberg, R. and Lerch, D. (eds.) (2010) *Post Carbon Reader: Managing the 21st Century's Sustainability Crises*, Healdsburg, CA: Watershed Media.

Hobson, K. (2006) "Bins, bulbs, and shower timers: On the 'techno-ethics' of sustainable living," *Ethics, Place and Environment: A Journal of Philosophy and Geography*, 9(3): 317–336.

Hopkins, R. (2011) *The Transition Companion: Making Your Community More Resilient in Uncertain Times*, Totnes, UK: Green Books.

Hornborg, A. (2011) *Global Ecology and Unequal Exchange: Fetishism in a Zero-Sum World*, London: Routledge.

Hult, A. (2013) "Swedish production of sustainable imaginaries in China," *Journal of Urban Technology*, 20(1): 77–94.

IEA (2008) *Promoting Energy Efficiency Investments: Case Studies in the Residential Sector*, Paris: OECD/IEA and AFD.

Jackson, T. (2009) *Prosperity without Growth: Economics for a Finite Planet*, London: Earthscan.

Karanja, N., Yeudall, F., Mbugua, S., Njenga, M., Prain, G., Cole, D. C., Webb, A. L., Sellen, D., Gore, C., and Levy, J. M. (2010) "Strengthening capacity for sustainable livelihoods and food security through urban agriculture among HIV and AIDS affected households in Nakuru, Kenya," *International Journal of Agricultural Sustainability*, 8(1–2), 40–53.

Keil, R. (2007) "Sustaining modernity, modernizing nature," in R. Krueger and D. Gibbs (eds.) *The Sustainable Development Paradox: Urban Political Economy in the United States*, New York: Guilford Press, pp. 41–65.

Knight, L. and Riggs, W. (2010) "Nourishing urbanism: A case for a new urban paradigm," *International Journal of Agricultural Sustainability*, 8(1–2): 116–126.

Kortright, R. and Wakefield, S. (2011) "Edible backyards: A qualitative study of household food growing and its contributions to food security," *Agriculture and Human Values*, 28: 39–53.

Kreuger, R. and Gibbs, D. (eds.) (2007) *The Sustainable Development Paradox: Urban Political Economy in the United States and Europe*, New York: Guilford Press.

Latour, B. (2004) *Politics of Nature: How to Bring the Sciences into Democracy*, Cambridge, MA: Harvard University Press.

Lin, W., Wu, T., Zhang, C., and Yu, T. (2011) "Carbon savings resulting from the cooling effect of green areas: A case study in Beijing," *Environmental Pollution*, 159: 2148–2154.

Merson, J., Attwater, R., Ampt, P., Wildeman, H., and Chapple, R. (2010) "The challenges to urban agriculture in the Sydney basin and lower Blue Mountains region of Australia," *International Journal of Agricultural Sustainability*, 8(1–2): 72–85.

Murphy, P. (2008) *Plan C: Community Survival Strategies for Peak Oil and Climate Change*, Gabriola Island, BC: New Society.

Naturvårdsverket (2008) *Konsumtionens klimatpåverkan*, Stockholm: Swedish Environmental Protection Agency (Naturvårdsverket).

Naturvårdverket (2012a) *Konsumtionsbaserade miljöindikatorer. Underlag för uppföljning av generationsmålet*, Stockholm: Swedish Environmental Protection Agency (Naturvårdsverket).

Naturvårdsverket (2012b) *Low-Carbon Transitions and the Good Life*, Stockholm: Swedish Environmental Protection Agency (Naturvårdsverket).

New Economics Foundation (2009) *The Happy Planet Index 2.0: Why Good Lives Don't Have to Cost the Earth*, London: NEF.

Parr, A. (2009) *Hijacking Sustainability*, Cambridge, MA: MIT Press.

Paul, A., Wiedmann, T., Barrett, J., Minx, K., Scott, K., Dawkins, E., Owen, A., Briggs, J., and Gray, I. (2010) *Introducing the Resources and Energy Analysis Programme (REAP)*, Stockholm: Stockholm Environment Institute.

Rockström, J. and Wijkman, A. (2011) *Den stora förnekelsen*, Stockholm: Medström.

Saar, M. (2010) "Stockholm mörkar sina verkliga utsläpp," *Effekt: Klimatmagasinet*, no. 2.

SEI (Stockholm Environment Institute) (2012) *Global miljöpåverkan och lokala fotavtryck. Analys av fyra svenska kommuners totala konsumtion*, Stockholm: SEI/Cogito.

Skovdahl, B. (2010) *Förlorad kontroll. Den ifrågasatta framstegstanken*, Stockholm: Dialogos.

Spaargaren, G. and Mol, A. P. J. (1992) "Sociology, environment and modernity: Ecological modernization as a theory of social change," *Society and Natural Resources*, 5(5): 323–345. Reprinted in A. P. J. Mol, D. A. Sonnenfeld, and G. Spaargaren (eds.) (2009) *The Ecological Modernization Reader: Environmental Reform in Theory and Practice*, London: Routledge, pp. 56–79.

Starke, L., Assadourian, E., and Renner, M. (eds.) (2012) *State of the World 2012: Moving toward Sustainable Prosperity*, Washington, DC: Island Press.

Stiglitz, J., Sen, A., and Fitoussi, J.-P. (2009) *Report by the Commission on the Measurement of Economic Performance and Social Progress*. Online, available at: www.stiglitz-sen-fitoussi.fr/en/index.htm (retrieved 1 July 2012).

Stockholms stad (2007) *Vision 2030: A World-Class Stockholm*, Stockholm: Stockholm City.

Stockholms stad (2010) *The Walkable City: Stockholm City Plan*, Stockholm: Stockholm stad.

Stockholms stad (2011) *Statistik om STHLM, Bostäder, Bostadsbyggandet 2010*, S2011:2, Stockholm: Stockholms stad. Online, available at: http://www.statistikom stockholm.se/attachments/article/65/S_2011_02.pdf (accessed 7 March 2013).

Stockholms stad (2012) *Stockholms miljöprogram 2012–2015*, Stockholm: Stockholm stad.

van Leeuwen, E., Nijkamp, P., and de Noronha Vaz, T. (2010) "The multifunctional use of urban greenspace," *International Journal of Agricultural Sustainability*, 8: 20–25.

van Veenhuizen, R. and Danso, G. (2007) *Profitability and Sustainability of Urban and Peri-urban Agriculture*, Rome: Food and Agriculture Organization.

Walker, P. A. (2005) "Political ecology: Where is the ecology?" *Progress in Human Geography*, 29(1): 73–82.

Wilkinson, R. G. and Pickett, K. (2010 [2009]) *The Spirit Level: Why Equality Is Better for Everyone*, London: Penguin.

Winner, L. (1980) "Do artifacts have politics?" *Daedalus*, 109(1): 121–136.

WEBSITES

Kairos Future (2010) www.kairosfutureclub.com/content/sveriges-milj%C3%B6export-g%C3%A5r-fram%C3%A5t (retrieved 5 June 2012).

Stockholms stad (2012a) http://bygg.stockholm.se/-/Alla-projekt/norra-djurgardsstaden/In-English/Stockholm-Royal-Seaport/ (retrieved 7 March 2013).

Stockholms stad (2012b) www.stockholm.se/OmStockholm/Vision-2030/Capital-of-Scandinavia/ (retrieved 5 June 2012).

Stockholms stad (2012c) http://international.stockholm.se/Future-Stockholm/Vision-2030/Innovation-and-growth/ (retrieved 5 June 2012).

Stockholm stad (2012d) *Klimatsmarta stockholmare*, www.stockholm.se/KlimatMiljo/klimatsmartastockholmare/ (retrieved 21 April 2012).

SymbioCity (2012) www.symbiocity.org/en/Concept/The-swedish-experience/ (retrieved 21 June 2012); see www.symbiocity.org for the new (2013) version.

CHAPTER 9
URBAN SUSTAINABLE DEVELOPMENT THE STOCKHOLM WAY

Amy Rader Olsson and Jonathan Metzger

9.1 INTRODUCTION

THE PRECEDING CHAPTERS have provided a broad introduction to some of the key issues, challenges, and knowledge areas in urban sustainable development, relating these questions to Stockholm's experience. This concluding chapter retraces some of the main themes of the book and asks what general lessons about urban sustainable development may be drawn from the specific case of Stockholm.

Knowledge is never eternal or neutral, but rather always irrevocably bound up within the wider societal conditions of its context of production. Half a century ago, KTH, the Royal Institute of Technology in Stockholm, published a summary of the then-current forefront of academic research, which on a wider level addressed the question, "How do we build a good city?" The authors who contributed to the anthology, *Bilstaden*, articulated this as a question of how to develop Stockholm into a modern city with vastly improved conditions for individual welfare through coordinated land-use and transportation planning (KTH, 1960). The anthology discussed many of the themes echoed in this volume: how to coordinate transport and land-use planning for high-quality workplace and residential development, how to balance central and suburban development, and how to work constructively to provide a balance of individual and collective goods and hence contribute to the general welfare. The difference was that in 1960 the solutions promoted by the KTH researchers all revolved around how to facilitate the "automobile city" (*bilstaden*), with "a large degree of individual freedom of mobility made possible by a departure from the idea of the traditionally centralized city, which to a large extent is determined by public transportation" (ibid., p. 7). The measures of success were also somewhat different from those of today, focusing almost exclusively on income and car ownership per capita.

Even though there is a world of difference between the Stockholm of today and the Stockholm of 1960, there are also important similarities: the city and its surrounding region are yet again grappling with the pressures and strains of rapid urban growth, such as bottlenecks and accumulating undercapacity in the housing sector and transport system. Whereas the 1960s policy response to these challenges facilitated a flight to the suburbs, current approaches instead focus on meeting the strong contemporary demand for distinctly urban amenities and centrally located housing and business locations.

The economy of the Stockholm region is currently the largest and most rapidly growing in Sweden. Despite a national culture still characterized by a pastoral romanticism, reminding us that this is historically quite a recently urbanized nation, the notion that Stockholm is an engine for Swedish national economic development is becoming widely accepted. Consequently, the discourse of development is changing, and is increasingly focused on global competitiveness and the urban agglomeration benefits of metropolitan regions. Further, the personal consumption of goods, services, and housing was both quantitatively and qualitatively different in 1960. Private consumption is now a major economic driver, in Sweden in general and Stockholm in particular, and this can be highly problematic from a sustainability perspective – no matter how "green" consumption is labeled as being.

The search for solutions to urban sustainability challenges attracts many foreign delegations to Stockholm, and the current generation of KTH researchers are still occupied with the question of how to build the "good city." Today, however, the response to this question reflects the revolution in ecological awareness and sensibility of the past few decades as well as the effects of an ongoing global economic recession, a global climate crisis, and looming resource crises. These fundamental insights provide the intellectual soil for conceptual innovations and areas of interest such as *ecological urbanism*, *active ground*, *green leases*, and *urban agriculture*, to name but a few of the emerging urban development concepts mentioned in the chapters of this book.

Nevertheless, the question remains as to what substantial and generalizable lessons can be drawn from Stockholm's specific experiences in its ambitions toward urban sustainable development. How much of Stockholm's development in this area is the result of knowledgeable and active decisions that could be replicated in other contexts, and how much is a consequence of natural endowments, geography and history? The next section of this chapter tentatively addresses this question and then provides a quick review of some of the specific particularities and blind spots of the current loosely sketched "Stockholm take" on sustainable urban development. We conclude with a presentation of a few conceptual pairings that, based on the content of this book, serve to highlight some of the difficult balancing acts and tensions that must be negotiated in any policy development scheme aiming at promoting urban sustainable development.

9.2 CREDIT WHERE CREDIT IS DUE: HOW DID STOCKHOLM GET HERE?

As is aptly illustrated by the different perspectives represented within this book, it is definitely possible both to applaud and to deride Stockholm's developmental decisions, both present and past. But perhaps more interesting and more valuable to urban professionals in other cities is to ask how Stockholm became what it is today. Have key policy decisions contributed to Stockholm's contemporary international reputation as a global beacon of sustainable urban development? Phrased more directly: is Stockholm's prizewinning urban environment merely the inevitable result of its natural and historical endowments in the form of its geographic setting, economic history, and cultural traditions – or can it also be traced to specific conscious and foresightful political decisions and investments?

Natural endowments

Sweden is both rather sparsely populated and comparatively rich in crucial natural resources, with an economy that has for decades been geared toward high-value-added exports. As such, this small country has managed remarkably well in a globalized world economy. Early and comprehensive investments in hydropower and nuclear power, although hardly feedstocks without environmental impact, nevertheless provide Stockholm with a stable source of low-carbon energy. Stockholm, having been a national capital for many centuries, is the central node in the national road and rail systems, and also enjoys a strategic maritime position between Lake Mälaren and the Baltic Sea, although the fact that Stockholm is built on numerous small islands across a sound also creates persistent challenges for local mobility. All that water also generates both stunning aesthetic values and practical hygienic amenities, with an almost unlimited supply of fresh drinking water (that is, as long as the brackish water of a potentially rising Baltic Sea can be kept out).

A short answer to the question posed above is thus that Stockholm's natural endowments and historical and geographical advantages have been important to the development of the city and have contributed to generating favorable preconditions for urban sustainable development policies, perhaps more so than in many other cities. But that said, we can also pinpoint important focused and conscious steps taken by Stockholm's politicians and civil servants in the preceding two decades to secure, manage, and develop the historical endowments that they have been entrusted with for present and future generations.

A mosaic of policies

The interesting questions now instead become: *which* policies appear to have made a real difference, *how*, and to *what effect*? These questions have been addressed in substantial detail in the chapters of this book, so here we can instead afford ourselves the opportunity to draw broad conclusions based on the insights provided in the preceding chapters. Roughly, we can divide these policies into three categories, which are also largely historically sequential.

Promoting human welfare and prosperity: This category includes sanitary improvements and transport investments that improved human life expectancy and preconditions for economic development, primarily instigated by liberal reformists around the turn of the last century with the broader purpose of increasing public welfare. These policies included the generally collectivistic and egalitarian public welfare solutions implemented under the auspices of the Social Democratic Party's welfare regime and the "People's Home" policy in the mid-twentieth century, stretching into the "Million (Homes) Program" for housing provision in the 1970s. They are also reflected in investments such as the extensive integration of public transport and other infrastructure with land-use planning, an approach manifested in suburbs such as Vällingby and Farsta and in district heating infrastructure. These policies provided practical solutions that improved basic human living conditions. Nonetheless, even if this was not their expressed aim, many of these interventions and investments also turned out to have unanticipated positive environmental effects. For instance, the extensive so-called green wedges between the suburbs and transport corridors today provide crucial ecological services, including recreational values for the city's inhabitants, support for biodiversity and species mobility, and carbon sequestration. Likewise, the high share of public transportation trips in the city, which keeps down pollutant emissions, owes much to the foresight of over-dimensioning the subway system during its construction between the 1940s and 1970s.

Environmental protection: In the late 1960s and 1970s, the state of the environment exploded as a central political topic, not only in Stockholm but all over the Western world. Fueled by NGO protests and activism, increasing ecological awareness led to demands for the legal protection of fragile ecosystems. Sweden was a forerunner and standard-bearer in the international arena, forging ahead with comparatively bold and broad national legislation for environmental protection, for instance the banning of a number of environmentally hazardous chemicals in the 1980s and 1990s. At the city and regional levels in Stockholm, spontaneous protests and strong NGOs had since the late 1960s put local environmental issues at the forefront of city politics, and achieved some substantial policy and plan revisions. Still, it was not until the 1990s and the Agenda 21 movement that environmental issues began to become a standing concern for the city administration.

Finally, in this context it would be remiss to pass over the latest generation of policies in this area – aiming at *sustainable city branding and marketing*. This set of policies, emerging locally in Stockholm roughly from the mid-1990s and onwards, positions Stockholm as an international best-practice example of urban sustainable development to promote not only the city itself but also a wide range of Swedish urban clean-technology products and services, particularly using the eco-district of Hammarby Sjöstad as a live showcase for these technologies and processes. While often generating concrete and substantial improvements in urban sustainability, these policies are heavily focused on those interventions that can be "packaged" and "put on display" to advance the reputation of the city and other public and private collaborators. This has prompted some critics to ask whether exportability and commercial potential are more important in the eco-districts of Stockholm than actual sustainability effects.

9.3 PARTICULARITIES AND BLIND SPOTS

Several cycles of policy intervention – some with environmental ambitions, others with unanticipated environmental benefits – have come together to generate the preconditions for Stockholm's strong development and reputation in the area of urban sustainability. But what are the specific particularities of the contemporary Stockholm "take" on urban sustainable development? And conversely, what important aspects are perhaps left out of Stockholm's approach – that is, what are its *blind spots*?

A culture of consensus – for better and for worse
Thomas J. Anton, professor of political science at the University of California, Berkeley, produced a major study of Stockholm's regional governance system in the early 1970s, resulting in the book *Governing Greater Stockholm* (Anton, 1975), in which he argues that Stockholm has had a unique political evolution. This would seem to leave little room for drawing lessons that might be applied elsewhere. However, he does note one thing that other places could learn from Stockholm: the masterfully executed techniques

employed in bringing together and generating agreements between politicians of different party affiliations, public servants, and professional experts,
as well as between different municipalities and between municipalities and
the national government.

This ability, which has come to be known as the *Swedish political
consensus culture*, is an amalgam of formal and informal techniques and
institutions for generating broad agreement in the Swedish political system
with respect to major development priorities (see also Chapter 2). It has been
widely praised both nationally and internationally as providing stability and
steady progress in Swedish society ("evolution rather than revolution"), but
has also been heavily criticized for marginalizing minority opinions and
interests – even raising claims that the Sweden of the mid-twentieth century
can be described as a corporatist society in which the political and business
elite, together with the strong unions, staked out the direction of societal
development without much real regard to fluctuations in public opinion. In the
Stockholm context, the consensus culture has played out as a highly informal
but quite stable pro-growth coalition between the biggest parties: the Social
Democrats and the conservative Moderates. Even if some major and contentious issues such as housing policy have been left out of this consensus,
and even if it has at times been forced to yield to pressure from NGOs and
public opinion, this informal political "meta-agreement" has nevertheless
proven remarkably robust and resilient over the decades.

In relation to urban sustainable development, one problem with this
stable agreement on broad development parameters has been the risk of
producing equally broad definitions of key concepts so as to avoid consensus-
threatening disagreements – with the result that these concepts become
highly superficial and thus unable to guide action in any substantial way.
Stockholm's city development policy promoting walkability, for example, is
often communicated as a major goal, but is not linked to specific indicators
or parameters that developers must meet. Stockholm promotes neighborhoods with a mix of ethnicities and socioeconomic backgrounds, but does
not mandate them in the way that some cities do, for example through
allocation rules for public housing. Thus, a risk inherent in the Swedish
consensus culture is that crucial differences are brushed over and important
decisions avoided, so as not to upset the consensus. In Stockholm, this
means that though all the major players are "for sustainability" as a key
development goal, a concrete operational definition of the concept is
repeatedly avoided so as to maintain consensus on the overarching goal.
When push comes to shove, the question of what sustainability substantially
entails and demands in the form of trade-offs and sacrifices is generally
avoided. When consensus becomes such a central goal in itself, it can
engender a very specific kind of decision-making sclerosis, despite broad
agreement concerning the key parameters of urban development.

Global, local — and the missing in-between

Stockholm's policymakers and research community would seem to be getting better at using models, methods, and processes that can analyze, predict, and to some degree engineer quite complex interactions in extreme detail. The Swedish high modernist intellectual tradition of systems analysis has found new expression in fine-grained analysis and solutions at the neighborhood and building levels. The chapters of this book highlight research insights related to the understanding of functional interactions — among people and firms in urban agoras (Chapter 5), integrated social-ecological systems (Chapter 4), closed-loop eco-districts (Chapters 4 and 7), between owners and tenants of green buildings (Chapter 6), and within the economic geography of metropolitan regions (Chapters 2 and 5). Eco-districts and their closed-loop resource cycling models are attracting worldwide attention owing to their ability to integrate both technological and management systems for water, sewage, and energy. The space syntax approach described in Chapter 4 and the urban agriculture trends reviewed in Chapter 8 are also good examples of neighborhood-level development based on an understanding of complex interdependencies between human and bio-physical systems.

Methodological and technological advancements supporting systems science have surely contributed to the focus on the fine-grained in many of the above-mentioned urban sustainability advances. Closed or semi-bounded systems at the smaller scale are easier to engineer and manage than the open-ended complexity of a whole city or urban region. At the national scale, the necessary level of abstraction makes systems approaches also fairly helpful as devices for thinking and planning with, although with large error margins and substantial degrees of uncertainty. But what of the middle scale? The focus on closed-loop local systems informed by fine-grained systems analysis, coupled with Stockholm's focus on international showcase neighborhoods, has led to a situation where the "broader picture" of the urban fabric becomes indistinct. For, notwithstanding the world-leading, cutting-edge eco-district developments of Hammarby Sjöstad and the Royal Seaport, what about everything around these areas? How, for instance, are we to retrofit and upgrade the large suburban housing estates of the Million Homes Program, with 200,000 apartments in Stockholm County (approximately 65,000 within city limits) in urgent need of renovation and improved sustainability performance? These make the city's designated eco-districts (Hammarby Sjöstad, the Royal Seaport, and Liljeholmen-Lövholmen), comprising a total of 26,500 apartments when fully completed in 2025, pale somewhat in comparison. And what about all the other new areas being constructed in the municipalities of the Greater Stockholm region but not designated as eco-districts, including the city's own major developments Västra Kungsholmen, Liljeholmskajen, and Hagastaden?

In relation to the sustainability-retrofitting of the Million Homes Program areas, there are promising experiments under way (see Chapter 8) – but these are nevertheless small-scale interventions, encompassing a few hundred apartments in total. Some argue that ecodistricts and small-scale demonstrations facilitate broader implementation of environmental innovations both locally and globally, but the question still remains as to the mechanisms by which this diffusion is to occur. Sustainable urban development, if implemented only in a piecemeal, market-led fashion at the neighborhood scale, may in the long run engender literally and/or symbolically secluded "premium ecological enclaves" (Hodson and Marvin, 2010) for a wealthy urban elite while failing to tackle the wider context of the surrounding urban fabric.

If we zoom out to the global scale, we find that Sweden is an active participant in international negotiations, regulations, and governance systems to protect global resources and ecosystems and that Stockholm has benefited from this stance. With an economy dependent on multinational enterprises and exports, and consumers dependent on imports, the Stockholm region's politicians and firms are keenly aware of the region's relation to the rest of Sweden, to the Baltic region, to Europe, and to the world. The general picture of Stockholm presented in the previous chapters is that of a metropole with well-developed mechanisms for global cooperation that help this medium-sized city in a small country act as a global player in sustainability issues – both in the public sector and within industry.

But between the global and national levels on the one hand, and the municipal levels on the other, there appears to be a lack of a stable "in-between" scale for efficiently dealing with urban sustainability issues – not least in Stockholm. The OECD (2006) has suggested that the Stockholm region's "weak middle" may be the result of metropolitan-level governance institutions that lack the legal authority to enforce land-use planning principles. This can be contrasted to Chapter 5 in this volume, which acknowledges the inherent conflict of interest among diverse municipalities, and between public and private actors, but sees a real potential for new metropolitan-level cooperative regimes and partnerships. At the broader regional level, the complexity of addressing both large- and small-scale change, slow and fast processes, and hundreds of goals and interests becomes overwhelming. Regional planners have responded by focusing more attention on subregional initiatives, acting successfully as mediators and coordinators – but perhaps again running the risk of losing sight of a bigger picture.

There is also a third missing "in-between" related to the relational geographies of sustainable urban development discussed in Chapter 8 – that is, the impacts and implications of action and consumption in one specific location on other local places, perhaps on the other side of the globe. Policymakers and the general public in Stockholm have a fairly robust comprehension of the connection between local activities and consumption and their possible detrimental global effects in the form of direct emissions of greenhouse gases that contribute to climate change. Nevertheless, there is still little general understanding of, and responsibility taken for, the ways in which the contemporary lifestyles of Stockholmers are fully dependent on an exporting of environmental "bads" to other places around the globe. Has Stockholm merely displaced its local emissions and pollutions rather than mitigating them? Stockholm is a metropolitan region whose population, trade, and growth dynamics are heavily influenced by – and highly influential upon – other regions and countries. It is no longer a cozy and somewhat isolated little hamlet at the edge of the extensive Scandinavian forest, but rather is part of the "global village," yet still with only a dim general understanding of its relation to the whole *ecumene* of life on the planet.

Urban social engineering in a cult of individualism?

The Swedish concept of a "people's home" (*folkhemmet*) was articulated in the late 1920s as the philosophy guiding the development of the "good society" that would give individuals and families the environment they needed to prosper. Even though the 2010s are radically different from those times, this very broad vision still permeates much of Swedish society and political culture – an ambitious welfare state striving to balance the needs of the individual with those of the collective. Several chapters in this volume highlight the challenge of not only maintaining these ideals but translating them into relevant principles in an ever more rapidly changing world.

History has taught us that the dividing line between individual responsibilities and rights and the sphere of collective decision making, institution building, and the provision of services is in no way set in stone, but rather is a fluctuating frontier determined by constant renegotiation based on the priorities and necessities of a specific society at a particular point in time. Chapters 7 and 8 argue that the contemporary strong and largely unquestioned focus on individual liberty, particularly in relation to personal consumption choices, may come at the cost of local and global sustainability. While the influential Swedish historical culture of the yeoman farmers, the *allmoge*, includes strong traditions of group consensus and community responsibility, the urban culture of consumption in the Western world in general – and major metropolitan regions in particular – has generally been described as highly individualistic and hedonistic. This tendency is further augmented in Stockholm by the fact that its population so far has been only mildly affected by the global financial recession of the early twenty-first century, and with a comparatively rich and young population, the personal consumption of resource-dependent products such as furniture, home appliances, cars, fashion, and luxury food items constantly reaches new record levels. With the highest share of single-person households in Europe (over 60 percent), housing prices are skyrocketing as individuals seek an urban lifestyle living in the studio and one-bedroom apartments that only a century before housed entire families.

Chapters 3 and 8 ask whether or not this current culture of consumption can be maintained in a city striving for sustainability. The right to express individual preferences through private consumption choices is something of the holy cow in Western late-modern liberal society, as evidenced by the storm of protest that erupted when it was suggested that such lifestyle choices might be regulated for residents in the new Royal Seaport eco-district (Chapter 7). But conversely, it may be that in an ever more individually based urban social framework, signals and banners of sustainability become important personal investments that accrue gains to both individuals and communities. In other words, conscious urban consumers may be willing to pay more for organic food with clear labels, to live in neighborhoods with "clean" reputations (such as eco-districts), and to use transportation modes marked as sustainable, so as to show that they are "part of the solution, not the problem."

Nevertheless, to base a sustainability strategy on visibility and prestige also poses dangers, for there are examples in Stockholm that indicate that sustainability issues with a low visibility factor may have difficulty in finding champions. The planned western ring orbital highway, the Stockholm Bypass, has been highly contested, in part owing to the projected increase in car traffic and greenhouse gas emissions associated with the project. However, when the project was redrawn and relocated mostly to tunnels passing under, instead of through, the attractive and wealthy areas rich in natural amenities on the outskirts of Stockholm, public discussion concerning the projected increase in greenhouse gas emissions became conspicuously muffled. Public opinion, even regarding invisible emissions, would seem to be dependent on visible manifestations of development and change. Even more worrying, the same highway link does in fact run overground next to some of the most socially and economically stigmatized neighborhoods of the city (those abutting Järvafältet), raising suspicions that even though enormous investments are made to keep the discomforts of increased car traffic out of sight of the relatively well-off, preserving their local landscapes, no such consideration is taken in less-well-off areas in the city.

As was discussed in Chapter 3, much current "sustainability talk" also builds upon the twin foundational notions that unlimited economic growth is possible within a sustainable future ("green growth") and that we can rely on individuals to make enlightened choices in their consumption if they are provided with sufficient information concerning the effects of those choices. But can we really put the future of the world in the hands of situated consumers, constantly burdened by economic, social, and practical pressures in their everyday choices? Chapter 3 raises concerns that individual pressures to save money, "keep up with the Joneses," or just grab something off the shelf in a hurry may increase the gap between stated environmental sustainability preferences and revealed preferences in consumption behavior.

From a certain perspective, investments in "clean tech," or green technologies, can avoid complete reliance on the situated consumer as the fundamental leverage point of sustainable development. Innovations such as smart windows, efficient water and sewage management systems, renewable energy, and energy management can establish frames of action that make sustainability outcomes less dependent on the everyday choice of individuals. These investments can to some degree, and with little drama, relieve people of the difficult daily choice between habitual comfort and making sacrifices (economic, social, practical) for the sake of "saving the planet." But when legislation does not require, or at least provide some form of reward for the use of, green building technologies, for example, those technologies must rely on accepted and established calculations of marginal cost advantages for their application – that is, they must "pay for themselves" by offering direct economic gains to developers and inhabitants in the form of energy savings, as exemplified in Chapter 6. Or they must be part of an eco-district or green building package, with tenants prepared to pay a little extra to be seen as "conscious consumers." This raises the question of whether ostensibly (but not necessarily substantially, as noted in Chapter 7) "sustainable lifestyles" are reserved for those with the extra cash to pay for the personal prestige afforded by owning an eco-district condo or choosing more "eco-friendly" or "biodynamic" goods and services. Should sustainable development be an *optional* choice for the *individual* few (and predominantly rich and well educated) or a *collective* endeavor paid for, managed, and executed collectively?

Stockholm's experience suggests that effective technical solutions for improving the sustainability of urban lifestyles must continually be balanced against the potentially issue-obfuscating and enthusiasm-dampening effects of passive sustainability technologies and infrastructure solutions. There appears to exist a need for constant reminders of the gravity of the current climate and environmental situation to individuals, firms, and society taken as a whole – particularly when effects are invisible or unfelt. Solutions that raise public awareness such as GlashusEtt, that provide economic incentives for more sustainable behavior such as the congestion charge (Chapter 7), or that facilitate new management or contracting arrangements such as green leases (Chapter 6) all appear to serve such a purpose.

The historical frontier between the private sphere of personal preferences and rights and genuinely collective concerns has been redrawn on many occasions, and perhaps we again need to think through how exactly to negotiate this divide as we confront a challenge to our survival as a species of unprecedented complexity that can never be fully addressed solely by individuals. Notwithstanding Sweden's international reputation for reliance on social engineering, for many decades the concept has primarily figured in the Swedish debate as a derogatory slur denoting undue governmental inter-ference in citizens' private lives and legal impediments to the free market. But perhaps the evolving climate crisis and other global environmental threats motivate some form of foresightful and democratically governed "social engineering" more than ever – that is, collective action to set necessary frames for individual choices. Clean-tech investments arguably function as a form of social (/technical/environmental) engineering aiming at securing sustainable behavior and lifestyles. Perhaps such an updated conceptualization of social engineering can remain open to include any tools that help both individuals and governments find democratically legitimate ways of reaching towards sustainable development in the city.

9.4 KEY STRATEGIC PARAMETERS FOR URBAN SUSTAINABLE DEVELOPMENT: LESSONS FROM STOCKHOLM

Those working toward achieving urban sustainability around the globe can find much inspiration in the experimental urban technologies, policies, and governance cultures of the Greater Stockholm area. But it is important not to let inspiration slip into seduction, for there are also some blind spots in the contemporary "take" on urban sustainable development currently being implemented in the city and region. Hopefully, this book has helped facilitate a fair and evenhanded appreciation of the strengths and weaknesses of current efforts toward urban sustainability in Stockholm. For there are definitely components of both hype and real substance in Stockholm's current reputation as a world-leading city in sustainable urban development. But what substantial lessons can be drawn from the case of Stockholm? The answer to this question most probably varies greatly for different readers in different local contexts. Thus, we will not conclude this book by offering some sort of marked score-sheet tallying up Stockholm's "sustainability score" compared to that of other cities and regions. Such simplifying summations can be important eye-openers, but they fail to convey an account that does justice to the entangled complexities of urban sustainable development.

Instead, we will conclude by building on Stockholm's experience, strengths, and blind spots in sustainable urban development to formulate a list of conceptual pairs that, taken together, suggest key strategic urban sustainable development parameters. The concepts in each pair are in no way mutually exclusive. Rather, they illustrate some crucial tensions, conflicts, or trade-offs in urban sustainability policies, as they encompass divergent but in a policy context often mutually reinforcing approaches – where ignoring one or the other term may obfuscate the importance of crucial angles.

Generating awareness or getting gritty

The City and region of Stockholm have spent considerable resources generating awareness of urban sustainability issues, both at home and abroad. And as Chapter 3 shows, the introduction of the environment as a topic of political discussion has had a tremendous impact on the decision-making climate, as has the contentious subject of climate change thereafter. Chapters 6 and 7 discuss the many efforts to raise public awareness regarding climate change, and Chapters 1 and 8 note the center-staging of sustainability issues in the city's current branding strategy.

At the same time, such information efforts always need to be balanced against – or, even better, combined with – concrete interventions that directly impact everyday life in the city. Such interventions might include the technological refinement of infrastructural systems, green construction, the creative use of pricing mechanisms or other economic incentives, or other new policies and tools to influence and affect the everyday habits of firms and urban denizens. While Stockholm's experience underscores the need to constantly communicate the challenges and potential of sustainable urban development, it also suggests that communication must be balanced with a comprehensive portfolio of concrete investments, regulations, and incentives. Likewise, action without communication may backfire if people do not understand the value of policies and investments.

Raising the bottom or shooting out peaks

Sweden's national approach to environmental protection has historically been to raise the bar on everything, using various forms of legislation to constantly push up the bottom and bring everyone up to par at a minimum level. Stockholm's local sustainability policy in the past two decades or so appears to have been almost the reverse: to maintain a low regulatory bar but show dramatic leaps over it – or, when the goals are more ambitious, to allow ample time to reach them or renegotiate them. As was highlighted in Chapters 4 and 8, international delegations flock to visit Stockholm's showcase projects and facilities, many of which demonstrate highly efficient, closed-loop, no-waste systems. In practice, however, these projects may be difficult to ramp up or out to other areas or contexts. This is partly a technological challenge, since showcases and pilots often use custom-designed or modified systems optimized to a limited site, building, or area. These may be better suited to high-budget urban brownfield redevelopments than to older public housing areas that were themselves, ironically, once the modernist showpieces that attracted those international delegations to the region.

So while there may be many marketing and branding advantages to be gained from "shooting out peaks" of sustainability performance, such as the construction of high-budget eco-districts, doing so may risk losing sight of the bigger picture: the rest of the city and all its inhabitants, and its relations to the surrounding world. It remains to be seen whether or not Stockholm's focus on showcases will facilitate the implementation of sustainable technologies and practices in more mainstream developments (as many claim) or contribute to the development of the fragmented and isolated urban gardens or eco-islands, as discussed in Chapters 5 and 8.

For sure, the City of Stockholm's new policy, demanding energy performance roughly equaling that of a *Passive House* (maximum of 55 kWh/m2) for all new developments on City-owned land (totaling approximately 70 percent of all land within City limits), maks an important and bold step in this direction and a foresigntful early implementation of the EU directive EPBD2.

Building consensus or forging ahead

Modern Swedish political culture has historically had a strong focus on consensual action, a firm agreement on central principles that allows for mobilization of collective resources. But as was argued in Chapter 3, broad consensus can also have a curiously paralyzing effect, for when it becomes of central importance to "get everyone on board," the risk is that concrete concepts and policies become so watered down that they are rendered practically inoperationalizable or extremely close to "business as usual," thus making them quite toothless as tools for change.

Admittedly, one of the strengths of the concept of sustainable development is that it promotes synergies in the form of solutions that can both secure the necessary resources for future generations and contribute to increased prosperity in the present. But in such a mind-bogglingly challenging quest as that which seeks to secure the preconditions for life on this planet, painful trade-offs are inevitable. Sustainability policies always generate winners and losers, and threaten powerful interests and regimes. Several chapters in this book seem to suggest that Stockholm could benefit from a reexamination of its own history, which at critical junctures managed to make bold policy and investment decisions while maintaining a firm focus on broad consensus regarding desired substantial outcomes.

Local benchmarking or a global outlook

As evinced by many of the chapters, including Chapters 1 and 8, Stockholm works locally to achieve urban sustainable development while proceeding from a fundamentally global outlook in its communication and marketing. What remain somewhat obfuscated in this geographical perspective are the concrete relations between places and the related displacements of environmental bads that occur by way of, for example, offshoring of production, as discussed in Chapter 8.

Greenhouse gas emissions and other forms of environmental damage do not disappear just because they are not produced in Stockholm. And it can be argued, as in Chapter 8, that if such emissions and environmental damage can be directly traced to the lifestyle choices and consumption practices of Stockholmers, they should also count toward the city's sustainability tally, which then becomes less impressive than currently claimed. Whether or not this is practically feasible, such a perspective highlights the fact that globally integrated urban areas support, hinder, and affect each other in complex ways. The lesson that Stockholm has yet to draw is that any effort to measure local policy impacts must proceed from a thorough understanding of interactions and influence chains that more often than not reach far beyond the borders of both the region and the country.

Durable, predictable frameworks or flexible, experimental solutions

Chapters 5 and 6 remind us that most built infrastructure is slow to change, and that this relatively fixed nature has lasting impacts – for better and for worse. It leads to stability of function over time, and helps individuals make decisions based on expectations that the road or bridge or airport or rail track will be there both today and tomorrow. Of course, such stable elements of the urban environment may facilitate more or less sustainable behavior, as discussed in Chapter 7. Even widely accepted sustainability measures – such as building a comprehensive metropolitan transit system – have in Stockholm contributed to a more dispersed and in some ways less energy-efficient settlement pattern, as described in Chapter 5.

Today, engineers and urban policymakers further recognize that adaptability and flexibility may be keys to the long-term success of major investments, and subsequently try to incorporate these properties into new infrastructure developments and institutional designs. But this is sometimes easier said than done. Some critics, for instance, argue that the ongoing construction of the Stockholm western bypass motorway locks the region into an unsustainable reliance on automobile transportation instead of facilitating a shift toward increased public transportation use. Others argue that without the bypass, the radial transportation corridor structure forces both public and private transport trips through ecologically and historically sensitive bottlenecks in the city center and isolates the northern and southern parts of the city – in other words, perpetuating a socially, economically, and ecologically unsustainable urban morphology.

Stockholm and Sweden are perhaps best known worldwide for large-scale investments in durable capital such as transit systems and the Million Homes Program. Nowadays, Stockholm's sustainability approach is more aptly characterized as a number of smaller-scale sustainability experiments and demonstration projects. These tend to be serious demonstrations, large enough to have measurable impacts and high visibility – such as the congestion charging experiment discussed in Chapter 7 – but still limited or tentative enough to guard against the risk that goes with implementing new technologies and policies. Historically, many of these demonstration projects have also been scaled up and expanded over time, for example the sewage management and district heating systems in Stockholm (Chapters 2 and 5).

Even if there are other drivers behind many of the contemporary experimental projects, as discussed in Chapters 1 and 8, Stockholm's emphasis on demonstration and pilot projects (which often become permanent) may nonetheless be an interesting component in any urban sustainability strategy if properly combined with broader measures and interventions. There are two important lessons to be drawn, therefore, from Stockholm's experience. The first is that the sustainability performance of durable capital infrastructure such as a new road can be radically affected by the short-term policies, incentives, and regulations regarding its use. The second is that progressive policies with strong sustainability potential, but clear winners and losers (such as congestion charging), may motivate experiments or demonstrations. Experiments and pilots can reduce uncertainty regarding effects and therefore allow policymakers to craft compensatory policies within the context of political negotiations.

Individual or collective action

In the past two decades, Stockholm's take on sustainable development has generally been market led, or at least market oriented, generating the necessary preconditions for market developments that favor sustainable outcomes. Examples include planning specifications requiring performance standards and supportive planning processes (as in the case of Hammarby Sjöstad), pricing and subsidy schemes (as with compost collection), or municipal procurement and sourcing policies favoring environmentally superior goods and services. Parallel to this market orientation, however, Stockholm also demonstrates reluctance to introduce new broad local legislation or planning regulations – with a few notable exceptions, such as the ban on studded snow tires along Hornsgatan in central Stockholm to reduce particulate emissions as well as the previously mentioned new energy standard for construction on City-owned land.

Market forces can have substantial transformative power, as manifested for example in the rapid increase in green building (Chapter 6), which is an entirely market-driven process. But nevertheless, Stockholm's experience illustrates that the market-based initiatives have limitations as a policy tool because they generally assume that individuals and firms are rational actors who can make informed judgments about the consequences of their choices. But taking into consideration all the various pressures that bear down on any individual in a specific situation of choice, reliance on the "enlightened consumer" may be shaky ground upon which to base a broad sustainability strategy. "Passive" sustainability technologies such as low-energy lightbulbs or smart windows take this responsibility away from individuals and thus create preconditions for more sustainable urban lifestyles that transcend the everyday choices of sometimes fickle consumers. But Stockholm's experience also indicates that passive solutions may engender apathetic consumers. If the lightbulbs are low-energy, why bother turning them off?

9.5 FINAL WORDS

Up to the end of the twentieth century, actions toward protecting the environment were generally considered to significantly improve quality of life through major innovations and investments in increased transit accessibility, cleaner water, and protected recreational and living environments. In more recent decades, and paralleling the rapid global diffusion of the Western consumerist lifestyle ideal, environmental measures have often been cast as threats or restrictions to the comforts of the "good life." The policies that have been most easily accepted are those that save consumers time, energy, or money as well as the environment. Recently, interest has turned to innovative and often "silent" or "passive" green technology, neatly avoiding uncomfortable disussions regarding established ideals of personal consumption, freedom and individual lifestyle choices. Nevertheless, if Stockholm is serious in its ambition to achieve sustainable urban development, the day may come sooner rather than later when some difficult choices have to be made. These concern not only how to ramp up innovations in technology and facilitate green consumption and green growth, but also how to balance individual and collective needs in the longer as well as the shorter term – taking into account the local, trans-local, and global impacts of local activities. In Stockholm, as elsewhere, this will require some form of broad common agreement, but an agreement that also grants a democratic mandate to experiment with both mundane and more spectacular policy interventions that impact how we live our individual lives.

REFERENCES

Anton, T. J. (1975) *Governing Greater Stockholm: A Study of Policy Development and System Change*, Berkeley: University of California Press for the Institute of Governmental Studies.

Hodson, M. and Marvin, S. (2010) "Urbanism in the Anthropocene: Ecological urbanism or premium ecological enclaves?" *City*, 14: 298–313.

KTH (1960) *Bilstaden*, Stockholm: Institutionen för Stadsbyggnad, Kungliga Tekniska Högskolan.

OECD (2006) *Territorial Review: Stockholm, Sweden*, Paris: OECD Publications.

INDEX

academic community 52
accessibility 11–12, 94, 107, 118, 121, 123
active ground 93–5, 196
activism 22, 198; Battle of the Elms protest f.38
actor-network theory 151–2, 155, 164
adaptability 209
AGA 19
agency 149, 151
Agenda 21 45, 51
agriculture 30, 185; modernization 15
Agyeman, J. 184
Ahrné, K. 83
air quality 28, 116; filtration 83–4; pollution 21–2, 35, 63
airtight building envelopes 133–4, 136
Akademiska hus 93, 95
Åkerman, J. 63, 65, 170
Albano University Resilient Campus project 93–7, f.96
Alberti, M. 83, 93
allotments 84, 185
American Apparel 184
analytical approach 82, 89–93, 97–8; sustainable cities 74, 82–8
Andersen, H. C. 14
Andersson, Å. E. 11, 41, 43, 105, 171
Andersson, E. 85, 93
Andersson, Magnus 20
Andersson, Monica 29
Andersson, O. 20
Andréasson, H. 162
Ängsö national park 57
Anshelm, J. 56
anthopocentricity 54
Anthropocene period 57
anti-inflationary measures 29
Anton, T. J. 199
applications in Swedish context 75–6
aquifer systems 113
archipelago challenge 115
architectural firms 188
Årets Stockholms byggnad 133
Arlanda airport 43, 113–14, f.114
Armstrong, D. 186
Association of Local Governments and Regions 140
Atlas 19
Atoms for Peace speech (Eisenhower) 35
automobile ownership and usage 25, 28, 35–7, 76, 196, 209 see also congestion charging
Axelsson, K. 47
Axelsson, S. 26
Ayres, R. U. 55

Backteman, O. 140
Baker, S. 52–4, 58
Balfors, B. 85
banking 19, 23
Barthel, S. 89, 93
BASTA 136
Bastiat, F. 23
Battle of the Elms protest f.38
Batty, M. 11, 90–1
Beatley, T. 74
Beckerman, W. 54–6, 58
Berdica, K. 110
Bernow, R. 90–1
Bettencourt, L. 108
betweenness analysis f.92
bilateral trade agreements 29
Bilstaden 195–6
biodiversity 82–3, 85–8, 185; Albano 95
biofuels 180, 191
biosphere 54
birth deficit 14
Björkman, L.-L. 84
Blå Jungrun estate, Parsta 133–5, f.134
black water 112–13
Bo01, Malmö 188, 190
Bodin, Ö. 85, 87
Bolund, P. 83
Bonde, M. 129, 139, 143
Borg, L. 143
Borgström, S. T. 85
Börjesson, M. 12, 117–19
Botsman, R. 179
Bradley, K. 178
Braudel, F. 10
BREEAM standard 130–1
Brembeck, H. 162
Bretton Woods 29
Brito, L. 74
Brotchie triangle 120–1, f.122
Brown, N. 131
Bruinsma, F. 117
Brundtland Report 45, 53–4
bumblebees 83
Byggvarudeklaration 136
Bylund, J. R. 152

Calthorpe, P. 74
Capital of Scandinavia claim 43, 171–2
carbon footprints 21, 177–9, 181, 183, f.182, t.178
Cars, G. 183
car-sharing 76
Carson, R. 22, 44
Central Board (stadskollegiet) 23
central railway artery (drawing, 1886) f.17
Central Station 32

certification systems 130–1, 133, 135–6
childcare system 39
Choay, F. 72
Choi, E. 91
circular resource flows 78 *see also* closed
    loop systems
city council 23
City Track Tunnel 116
city-nature dichotomy 88–9, 94
civil society 20, 60
class issues 20, 29
classification systems 129–31, 133
clean-tech industry 176, 205–6
climate change 44, 47, 84, 95, 110, 113,
    184, 207
climate debt 179
climate positive developments 81
closed loop systems 113, 201, f.114
cloud computing 115
club goods 107
coal 28
cognitive level of urban space 90–1
Colding, J. 89, 93–4
collaborations 93
collaborative consumption 179
collectivism 203, 205, 210–11
colonialism, opposition to 38
commercial green buildings 131–3, 140,
    f.132
commodification 175
communications 25
commuting 33, 44
competitions 133
competitive uncertainty 110
conflicting goals 52, 62–3, 67, 84–5
congestion charging 45–6, 116–17, 125,
    148–9, 160–4, 205, 209, f.160
connectivity 83, 85, 93
Connelly, S. 52, 58
consensus 40, 53, 60, 66–7, 159, 175,
    199–200, 203, 208
conservatism 15
Constantinople, comparison with 14
construction: costs 137; materials 136;
    waste 34
consumer goods 15, 18–19, 27, 29, 35
consumption 27–30, 80, 196; culture of
    203–5; minimization of 78
consumption perspective 177–9, 188–9,
    203, 209
contractual innovation 142–3
Cordell, D. 185
Corner, J. 75
corporate capitalism 175
cosmopolitanism 42–4
Costanza, R. 83
cost-benefit analyses 118–19
County Administrative Board 33, 44, 85
Covenant of Mayors 2007 45
creative sector 172
credit systems 29

crowding 18
cruise ships 81
cultural ammenities 76, 81, 172
cultural contexts 54, 98, 126, 150–1
Cutini, V. 90
cycling 18, 36, 125, 170–1
Czarniawska, B. 152, 155

Dagens Nyheter 147, 157, 159
Daily, G. C. 57
Danso, G. 186
Darley, J. 184
data-driven sustainability 116
decentralization 80, 104
decision-avoidance 200
decision-making 51–3, 105, 110
decoupling 177–8, 188, 190
defense 23
democracy 15, 23, 25, 60, 105, 206, 211;
    e-democracy 115, 125
demonstration projects 124, 162, 164,
    188, 201–2, 207–10
Dennis Package of transport investments
    43
densification 16–18, 43, 91, 93, 125
depoliticization of environmental issues
    52–3, 61, 66–7, 190
design theory 89, 97–8
Desyllas, J. 90
Deutsch, L. 55
devaluation 39
developing countries 53
development: directions f.104;
    endogenous/exogenous 22
Diaz-Jernberg, J. 137
discursive practice 149–52, 155, 163–4,
    175
discursive struggles 52–3, 152, 164
distance-based systems 118
district heating 35, 45, 110, 114, 124–5,
    198, 209
Dittmar, H. 76
diversification 24–5
diversity 42; ethnic 30, 33, 42, 200;
    transportation 76 *see also*
    biodiversity
Djurgården 170
domestic migration 11–12, 26–7, 44
Droege, P. 74
Dryzek, J. S. 60, 175
Duany, A. 76
Dufwa, A. 25
dynamic mechanisms: ecosystems 88;
    infrastructure 108–11, f.109

eco-branded products 169, 174–5, 178–9,
    188, 190, 199, 205, 207
eco-cycle model 78
eco-districts 199, 201, 204–5, 207–8
eco-fascism rhetoric 147
ecological connectivity zones 94

ecological footprints 21, 177–9, 181, 183, f.182, t.178
ecological modernization 56, 62, 169, 175–6
ecological urbanism 72–3, 75–6, 82, 196
economic crises 24–6, 39–40, 170, 184, 203
economic efficiency 64–5
economic geography 201
economic growth 15–16, 29, 42, 46, 48, 59–61, 175, 177, 196; conflict with environmental protection 56–7; crisis (post-1970) 38–42; green 204; as paradigm 170, 175, 184; Smart 76; sustainable 51, 62, 176–7
economic policies 39, 45
economic studies of green buildings: existing 140–3; new 136–40
economies of scale and scope 108, 114
ecosystem design: predictive tools 85–7; in urban design 88–9
ecosystem dynamics 88
ecosystem-human relations 57, 84–5
ecosystems approach 51, 54, 56, 82
ecosystems services 83–4
e-democracy 115, 125
education 148, 152, 154–6, 186, 207
Eichholtz, P. 137
Eisenhower, D. D. 35
Ekins, P. 55
electric vehicles 116
electricity sector 45
electricity usage 28; increase 35
electrification 19, 22, 25
electronics industry 41
elitism 80, 202, 204
Elmqvist, T. 83
emissions 62–4, 116, 144, 176, 204, 209; automobiles 36; calculation methods 177–9, t.178; decreased 47, 76, 188; imported 177, 203; increased 171, 177–9, 181, 189; particulate 36, 61, 116, 210; underestimation of 65 see also consumption perspective
energy 35; from waste 34
energy efficiency 47, 75–6, 125; green buildings 129, 132–6, 140–1, 144, 159
Energy Performance Contracting 143
energy sector 45, 197
energy systems 113–15, f.114
engagement 151, 154
Engineered Garments 184
engines of transition 116
entrepreneurship 42, 175
environmental balance: splits in 26–9
environmental cost of modernity 21–2
environmental effects exported 28
environmental impact assessment 87–8
environmental issues: multiple definitions of 51–2
environmental justice 57–8, 63, 176–88

environmental policy 38–9, 44, 47, 59
environmental programs 45, 59–60, 66, 154, 178
environmental protection 43, 81, 198–9, 207, 211; conflict with economic growth 56–7
Environmental Protection Agency 22, 177, 179, 181
Environmental Quality Objectives 61–2, 66
environmental rating tools 130–1
epistemological framework, design theory as 97–8
equality policies 39
Ericsson group 41
Eriksdal waterworks f.21
Eriksen-Hamel, N. 186
Eriksson, E. 20, 29
Ernstson, H. 83, 93
essentially contested concepts 58
ethanol 180
ethnic diversity 30, 33, 42, 200
Eurasian lynx 85
European Commission 46
European Union 45, 87, 130, 133
export orientation 29, 40; planning models 80
extreme weather events 84

Faeth, S. H. 186
Faith-Ell, C. 143
Farr, D. 74
Farsta 198; nuclear reactor 35
fast/slow processes (infrastructure dynamics) 108–11, f.109
FEBY standard 133, 135
ferries 18, 81
Finnveden, G. 63, 65, 170
Fisher, D. R. 87, 175
flexibility 209
flooding 113
Florida, R. 183
Folke, C. 89
food miles/security 186
foreign aid 38
Fortum 152, 154
fossil fuels 35–6, 40, 45, 61, 114, 180, 185, f.173
Foucault, M. 158, 164
fourth logistical revolution 41
free trade 16, 29
Freudenberg, W. R. 175
Fröberg, L. 139
functional properties of infrastructure 107
Fürst, F. 120–1

Galaz, V. 57
Galli, E. 180
garbage collection 114–15
garden cities 24
garden city influence 20

gardening 84
gas production 20–1, 45
GATT 29
GDP growth 15, 30, 42, 177, 180–1, 183, 188, 190, f.182 *see also* economic growth
GDP-CO$_2$ relationship 188
gender issues 23–4, 39
geographical context 103, 105, 115, 197, f.104
geographical information systems 85–6, 89, 91–2
geothermal energy 114
Gibbs, D. 174–5, 188
Gill, S. E. 84
GlashusEtt 148, 152–6, 163, 205; poster f.153
Glasson, J. 87
global trade 29, 197
global urban system 11–14
globalization 169, 184
globalized service society 183–5
global-local perspective 75, 201–3, 208–9
goal conflicts 67
Gontier, M. 86
Goossens, Y. 180
Gothenburg 41
governance 23, 39, 46–7, 78, 149, 158, 199, 202
governmentality 158
Graham, S. 108
Grahn, A. 142
Great Depression 25–6
Green, A. 154
green arteries/corridors 84–5, 93–5, 107
Green Building Council 130
green buildings 129–44, 201, 210; commercial/public sector 131–3, 140, f.132; economic studies 136–43; environmental rating tools 130–1; existing buildings 140–3; innovation diffusion in construction 135–6; new buildings 136–40; residential 133–5, 141, f.134
Green Capital of Europe, 2010 10, 46, 169, 176
green city concept 43
green economy 188
green growth 204
green leases 132–3, 142–3, 196, 205
Green Party 59
green roofs 76, 94, 132
green spaces 63, 82–3, 185–7, 198; biodiversity 83–4, 88; ecosystem-human conflict 84–5
green technology 180, 205
green walls 94
green welfare state 56
Greene, M. 90
greenhouse gases 47, 63–5
greening of existing buildings 140–3

Grewal, P. S. 186
Grewal, S. S. 186
grid principle 20
Gripenstedt, J. A. 23
Gröna Folkhemmet 56
Groth, K. 78–9
ground heat extraction f.114
groundwater: heating/cooling 113–14; salinization 115
guilds' monopoly repealed 18
Gullberg, A. 26, 33, 124
Gunnarsson-Östling, U. 59–61
Gustafsson, T. 147

Haas, T. 73–4, 76
habitats 83, 85, 88
Hagastaden 201
Hägerstrand, T. 107, 123
Hagman, O. 162
Hajer, M. A. 56
Hall, P. 31
Hållbara Järva project 187
Hallerdt, B. 28, 35
Hammarby Sjöstad 43, 71–2, 76–81, 148, 152, 154, 156, 163–4, 171–2, 174, 188, 190, 201, 210, f.77
Hammarström, I. 18–19
Hanson, J. 89–90
Hansson, S. O. 55
Happy Planet Index 181
harbor expansion 18
Hårsman, B. 13, 126
Harvey, D. 53, 57, 176
health 16, 20–1, 30, 159
heat island effect 12, 47
heat recovery systems 133, 136
heating, district 35, 45, 110, 114, 124–5, 198, 210
heating from power plants 35, 40
heating/cooling management 113–14, f.114
Hedrén, J. 56
Heinberg, R. 180, 184
Henderson, J. V. 12
Henriksson, G. 161–2
Hernes, T. 155
heterogeneity 93
hill parks 16–18, 34
Hillier, B. 89–90, 97
Hirdman, Y. 159
historical perspective 10–48; mid-19th century to World War I 14–23; World War I to World War II 23–9; World War II to the 1970s 29–38; 1970s onwards 38–46; global urban system 11–14; past/future comparison 46–8
Hobson, K. 175
Hodson, M. 202
Högberg, L. 141
holistic approach 74–5, 78
Holling, C. S. 55, 89

Hopkins, R. 170, 184
Hornborg, A. 180
housing provision 34, 127; funding 33;
    green buildings 133–5, 141, f.134;
    policy 20, 24; shortages 12, 18
Hudson, W. R. 106
human capital 55
Human Development Index 180
human needs 53–4; satisfaction of 55–6
Hunhammar, S. 83
Hunt, D. 51
hydroelectricity 28, 35, 40
hygiene standards 21

IBM 41
ICT 42, 115–16
ICT-electronics clusters 41–2
Iida, S. 90
IMF 29
immigration 30, 33, 44; poster (1946) f.27
imports/exports 4, 14, 16, 18, 23, 28,
    39–40, 190; and emissions 177, 203
incentives 142–3, 148, 163, 205
incremental provision of infrastructure
    108
individual metering 142
individualism 38, 203–6, 210–11
indivisibility of infrastructure 107
industrial evolution 18–20
industrial failures 39
industrial growth 29–30
industrial policy 41, 43
industrial suburbs 18
inequalities 12, 57, 176
information campaigns 148, 152, 154–6,
    186, 207
information technology 41
infrastructural evolution 16–18
infrastructure capital t.106
infrastructure development and planning
    102–27, 209–10; challenges 111–20;
    definitions 105–11; sustainable
    scenarios 120–4
infrastructure dynamics (fast/slow
    processes) 108–11, f.109
infrastructure-led scenarios t.111
Ingo, S. 34, 43
inner city 16, 18, 93
innovation diffusion, green buildings
    135–6
institutions: performing sustainability
    149–52
insulation 133–4, 136
integrated systems approach 78–9, 201
integration-isolation nexus f.123
intellectual capital 48
inter-agency cooperation 82
interdependencies 13–14, 45, 110, 176,
    183, 201
interdisciplinary approach 78, 95
internalization of costs 56

International Atomic Energy Agency 35
international branding 174
international corporations 19
international sustainability 176
internet 115
intragenerational justice 57–8
investor perspectives 129, 139
iron ore 16, 23, 39
Isaksson, K. 46

Jackson, T. 184
Jacobs, M. 58
Jaffee, D. 138
Järvafältet 204
Jenelius, E. 110
JM 157, 159
job creation 26
Johannisson, K. 159
Johansson, B. 24, 42, 105, 107
Jonsson, N. 142

Kallstenius, P. 20
Källström, U. 157
Karanja, N. 186
Karlström, A. 92
Kay, A. 151
Keil, R. 176
Kennedy, C. 84
Kenworthy, J. 53, 120
Keynesian approach 25
Kista district 41–2
Klara pit f.32
Knight, L. 186
knowledge: creation 13, 195; exchange
    72, 93; networks 126; systematic 73
knowledge-intensive industries 25, 40–2,
    72, 183
Koch, D. 90–2
Kortright, R. 186
Kreuger, R. 174–5, 188
KTH (Kungliga Tekniska högskolan) 8, 19,
    22, 35, 72, 77–9, 89–92, 97, 129–30,
    181, 195–6
Kuiken, H. 139

L. M. Ericsson 19
labor market 19, 23–5, 39, 42
labor-state relations 25–6, 29, 41
Lafferty, W. M. 52
Lake Bornsjön 112
Lake Mälaren 103, 112–13, 197
land ownership 18
land use patterns 118–19, f.120;
    integration-isolation nexus f.123
Landing, G. 159
landscape connectivity 90, f.86
landscape urbanism 75–6, 82, 85
large-scale transport flows 13
Larsson, Y. 23, 26
Latour, B. 151, 164, 176
Law, J. 151

Lawson, B. 97
Le Corbusier 97
leadership functions 37, 76, 80
LEED standard 130–1, 133, 135
Left Party 59
Legeby, A. 90–1
leisure 185
Lélé, S. 58
Lerch, D. 184
liberal capitalism 15
Liberal Party 59
liberal reforms 23
life expectancy 21, 30, 46
Life Satisfaction Indexes 181
life sciences 41
lifestyle choices 154, 158–9, 164–5, 183,
     189–90, 203–5, 210
Lilja, S. 51, 58
Liljeholmen-Lövholmen 201
Liljeholmskajen 201
Lin, W. 186
Lind, H. 141, 143
linear consumption flows 185
linear development 180
Linzie, J. 42
Ljung, B. 59
localized economies 19, 183–5 see also
     global-local perspective
logic of practice 151
Lombardi, D. R. 51
Lundewall, P. 18, 20
Lundqvist, L. 121

Magnusson, L. 19, 23, 31
Mahapatra, K. 141
maintenance costs: green buildings 138
Mälardalsrådet 48
Malmö 41, 188
management and development functions
     31
Mandell, S. 139
Maneekum, F. 140
manufactured capital 55–6
manufacturing 16, 18–19, 28, 30–1, 41–2,
     184–5
Marcus, L. 89, 91–4, 97
marginalization of minority opinions 200
market perspective 59, 80, 168–71, 188,
     190, 199, 202, 210; eco-branded
     products 174–5, 178–9, 188–90, 205,
     207
Marshall Plan 29
Marvin, S. 108, 202
Master Plan (1952) 32
MatrixGreen 87, 90, 94, f.86
Mattsson, L. G. 110, 117
Meadowcroft, J. 52–3
medical research 41
Meffe, G. K. 55
Merson, J. 186
micro-climate regulation 83

micro/macro levels of urban processes
     108–11, f.109
micro-technology 41
migration 26
Miljöbyggnad (Environmental Building)
     system 130–1, 135–6
Millennium Ecosystem Assessment 83
Million Homes Program 30, 141, 198,
     201–2, 209, f.33
Ministry of Finance 12
Mitcham, C. 58
mobility behavior 120–4
Moderate Party 59, 200
modernism 29
modernity: environmental cost 21–2
modernization 24–5; agriculture 15;
     ecological 56, 62, 169, 175–6
Mokhtarian, P. 123
Mol, A. P. J. 175
Monocle 184–5
Mörtberg, U. 83, 85
Mostafavi, M. 75
motorways 64–5, 170, 209
Mouffe, C. 175
multiculturalism 30
multifamily housing 24, 141
Munda, G. 55
municipalities 30
Murphy, P. 184

nanoparticles 116
national parks 57
natural capital 54–6, 59, 61–2
natural resources 16, 23, 53, 180, 185,
     190, 197
nature reserves 34
nature-culture symbiosis 13
Naturvårdsverket 22, 177, 179, 181
NCC 135
neighborhood unit model 71–2, 75, 201
networks 11, 37, 107, 126 see also
     actor-network theory; transit network
Netzell, O. 90
Neumayer, E. 55
neutrality in wartime 23–4
New Economics Foundation 181
New Urbanism 75–6, 88
Newman, P. 53, 120
Nobel 19
Nordic Swan Ecolabel 131, 135–6
Norrvatten Association 112
nuclear power 35, 40
nutrient leakage 115

Ödman, L. 142
OECD 124, 202
Ohland, G. 76
Ohlsson, P. T. 23
Ohlsson, S. 138
oil 36, 40, 180
Olofsson, B. 115

Olsson, K. 109
Olsson, A. R. 126
1 percent goal for foreign aid 38
open production 185
operating costs: green buildings 138
operationalization of sustainability
    concept 52–3
oppression, opposition to 38
orbital links 34, 64–5, 116
organizational relocation 39
Outline Regional Plan (1966) 33–4
owner-occupiers 141

Palazzo, D. 74
paper pulp industry 35
parking fees/spaces 116, 159
Parr, A. 175
participation approaches 60
particulate emissions 36, 61, 116, 210
passive solutions/technologies 129,
    133–7, 156, 210–11
passivization of citizens 159
path dependency 149–52, 164
Paul, A. 179
peak everything 180
Pearce, D. W. 55
pedestrian movement 90
peer-to-peer sharing 179
Pennfäktaren 11 commercial building
    131–3, f.132
performativity 93–4, 149, 151–5, 164
peri-urban districts 83
personal travel 162, f.117
Persson, G. 56
petroleum 35–6
Pettersson, R. 21, 28
Pickett, K. 181, 183
Pierson, P. 151
piped water systems 20
Place Syntax Tool 91–2, 94, f.92
place-making 75
planners 51, 67
planning 31–2, 43, 52–3, 66–7, 105, 191;
    EIA and SEA requirements 87–8;
    infrastructure 108–10, 112, 126–7;
    and market outcomes 120–4
Planning and Building Act (1987) 43
polarization 175
policy instruments 148–9, 156, 160, 164,
    198–9, 207
political ecology perspective 168–9,
    174–6, 183, 190
political evolution 15–16, 39, 43,
    199–200
political factors 25–6, 125
pollination 83
population: decline 33, 39; growth 15–16,
    24, 26, 33, 44, 46
Porter, L. 51
possibility, theories of 72
postwar renewal 29–34

power plants 28, 35
power relations 176, 183
precautionary principle 110
predictive tools, urban planning 85–7
preservationist approaches 33, 56–7
private landlords 140, 142–3
private sector employment 39
private vs. public provision 108–10, 114,
    126
private wells 115
private-public partnerships 143, 174, 191
privatization 45, 126
productive landscapes 185–7
professional practice 91
profitability 42, 139–41, 175
Promenadstaden 52, 173, 184
Property Owners Federation 140
property values 137, 139–40
protection of environment 43, 56–7, 81,
    198–9, 207, 211
protest 38
public costs of urbanization 12
public elevators 18
public goods 107
public opinion 22, 36, 39, 43, 163, 200,
    204
public parks 16–18
public transport 45, 71, 116, 125, 162,
    170, 190, 198, 209, f.117
public vs. private provision 108–10, 114,
    126
public-private partnerships 143, 174, 191

quality of life 181, 183
quasi-static uncertainty 110, 115
Quigley, J. M. 13

railways 16, 35–6, 46
rainwater drainage 83
Ranhagen, U. 77–9
rationing 28–9
real estate 20, 129, 171 see also green
    buildings
rebound effects 123
recycling 20, 34, 76, 80, 132, 178–9
Redclift, M. 53–4
Reduced Climate Impact objective 64
Rees, W. E. 55
refugees 30, 33
regional administration 24, 202
regional development 39, 41
regional policy measures 12
regionalization 172
relocation of national organizations 39
renewable energy 45, 76, 78
renovation 131–2, 141
rental income: green buildings 137–8
repertoires 97
research and development (R&D) 37, 42
research institutes 41
resident amenities 76

residential green buildings 133–5, 141, f.134
resilience theory 89, 93
resources debate 26
responsible lifestyles 154–60, 164, 178, 189, 203, 205, 210
Retelius, A. 138
retrofitting 187–8, 190, 201–2
reuse 78
Reuterskiöld, A. 139
Richert, G. 22
Rietveld, P. 117
Riggs, W. 186
Rio Conference (1992) 45, 51, 60
risk 138, 209
road pricing systems 116–18
roads 34, 36, 64–5, 116, 204
Robinson, J. 53, 58
Rockström, J. 57, 184
Rogers, R. 179
role model, Stockholm as 25–6, 169, 171, 187–8
Rolston, H. 54
Rönkä, E. 115
ROT Award 133
Royal Institute of Technology 8, 19, 22, 35, 72, 77–9, 89–92, 97, 129–30, 181, 195–6
Royal National City Park 80–1, 93, 107
Royal Seaport 43, 80, 155–6, 164, 171–2, 174, 201, 204, f.81; performing sustainability 147–8, 152, 156–60; perspective sketch f.157; urban design 80–2
Rutherford, J. 80

Saar, M. 177
Sandgärde, M. 137
sanitation 16, 20–1, 28
satisfying preferences 55–6
SBoxes 134
Schéele, S. 26
Schön, D. 97
Schumpeter, J. A. 41
scientific advances 19, 59
sea levels 103
Separator 19
service industries 30–1, 37, 40, 172, 183–5
settlement and workplace patterns 120–2, 124–5, 209
sewage systems 22, 28, 34, 76, 112–13, 210
shipyards 39, 41
showcase focus 207–8
Shrader-Frechette, K. 57
Sidenbladh, G. 33
Simon, H. 72
Skanska 133, 135
Skovdahl, B. 180
slums 12

Slussen 103, 113
small-scale catalytic interventions 74
Smart Growth 76
Snickars, F. 26, 105, 107
social construction of needs 54
Social Democratic Party 20, 25–6, 39, 46–7, 56, 159, 198, 200
social engineering 47, 147–8, 203–6
social homogeneity 71–2
social justice 177
social segregation 170
social-ecological systems 57, 89, 93–5
societal formations 183–4, f.184
socioeconomic development 53, 56, 186
socioeconomic diversity 200
socioenvironmental impacts 179
socioenvironmental movements 170
sociomateriality 149–52, 164–5
soil remediation 81
solar panels 76, 132
Soviet bloc 29
Spaargaren, G. 175
space syntax 89–91, 201
spaces of knowledge exchange 72, 93
spatial capital 91–3
spatial form, systematic knowledge about 73
spatial morphology 89–93
spatial performativity 95
spatial properties of infrastructure 107
species migration 83
speculative theory 72, 74–5, 82, 97–8
Stafford-Smith, M. 74
Ståhle, A. 90–2
Starke, L. 181
steam power 19, 21
steel industry 35, 39
Steen, J. 91
Steiner, D. 74
Stiglitz, J. 181
Stockholm Agreement 63–4
Stockholm Award for Building of the Year 133
Stockholm Bypass 64–5, 170, 204, 209
Stockholm Chamber of Commerce 185
Stockholm Conference (1972) 38, 44
Stockholm County Council 33, 41, 43, 45, 58–9, 66
Stockholm Declaration (1972) 38
Stockholm Environment Institute 179
Stockholm Water Company 154
Stone, B. 12
Strandh, S. 22
strategic alliances 126
strategic environmental assessment 87–8
streetcars 18, 28, 35
Strömqvist, U. 42, 105
strong sustainability 61–2
structural challenges 170
substitutability of different types of capital 55–6

suburban garden scenario 124
subway system 12, 26, 32, 36–7, 76,
    118–19; and use patterns f.120
SundaHus 136
Sundborn, D. 139
sustainability requirements 81–2
sustainable cities model 188, 199;
    analytical theories of 82–8; Hammarby
    Sjöstad 77–80
sustainable development concept 51–67,
    173, 175; planning 62–5; political
    discourse 58–62
sustainable growth 51, 62, 176–7
sustainable prosperity 181, 183, 190
sustainable urbanism 74–6, 79, 82, 88,
    168, 171, 174, 202
Svane, Ö. 154
Svenska Bostäder 133, 135
Sweco 77, 79
SymbioCity 77–80, 82–3, 174, 188–9,
    f.79; Swedish Experience promotion
    graph f.189
synergies 79, 208
system-and-network properties 107
systems analysis 120, 201

tariff reductions 29
taxation 39, 45–6, 172
technical properties of infrastructure
    105–6, t.106
techno-industrial progress 56
technological innovation 19, 41, 48,
    60, 124, 135–6, 158, 172–4, 190–1,
    211
telecommunications systems 115–16,
    123, 126
TellHus 135
temporal properties of infrastructure
    106–7
tenant satisfaction: green buildings 137–8
tenant-landlord relations 142–3
Tengbom Architects 154
territorialism 177–9
The Walkable City 52, 62–4, 66, 173, 184
theories of possibility 72
theory in art 97
theory-based professionalism 73
Thisse, J.-F. 12
tipping points 108–9
Toller, S. 144
Törnqvist, G. 108
Tottmar, M. 147
tourism 42, 170
toxins 44
trade 14, 23, 25
trade-offs 53, 66, 80, 123, 200, 208
traffic problems 33, 45, 47, 117–18, 162
    see also congestion charging; parking
    fees/spaces
trams 76
transit network 36, 46, 84, 196, 209

Transit Oriented Development 76
Transition movement 170
transponder system 161
transport infrastructure 52, 63–5;
    complexities of sustainable 116–20;
    user behavior and welfare
    development f.122
transport system 25–6, 28, 43, 60–1, 63,
    103–4; costs 121, 123; diversity 76
travel reduction 115–16, 119, 123
trees 81

uncertainties 110–11, 115, 122, 210
unemployment 26
unions 19, 26
United Kingdom 130, 142
United Nations 38, 53, 60, 65, 130
United Nations Environment Programme
    38–9
United States 57–8, 130, 135
universities 41
urban agora scenario 124, 201
urban agriculture 84, 88, 185–7, 196, 201
urban densification 16–18
urban design theory 71–98; Albano
    Campus 93–7; as an epistemological
    framework 97–8; analytical theories of
    sustainable cities 82–8; analytically
    supported theory 88–93;
    contemporary trends 73–6;
    Hammarby Sjöstad 77–80; Royal
    Seaport 80–2; social-ecological 93–4
urban green areas see green spaces
urban metabolism 84
urban morphology 88–91, 93–5, 209
urban planning predictive tools 85–7
urban re/development 31–3
urban subsystems 78
urbanization 11–12, 24

Valeskog, J. 159
Vällingby 43, 71–2, 76, 198, f.31
van der Schaaf, K. 137
van Leeuwen, E. 186
van Veenhuizen, R. 186
Vasakronan 131–2, 142
Västra Kungsholmen 201
Vaughan, L. 90
Veidekke 135
ventilation 136
versatility in use 107
videoconferencing 116
visibility factor 204
Vision 2030 project 60–4, 66, 171–3, 184;
    document excerpt f.173
voluntarism 159, 178

Wackernagel, M. 55
Wakefield, S. 186
Waldheim, C. 75
Walker, P. A. 176

wall thickness 136
Wallace, N. 138
Wallenberg, A. O. 23
waste management 20, 34, 76, 80,
    114–15
wastewater 112–13, 115
water: consumption 134; levels 103, 113;
    purification 22, 80, 112; quality 28, 34;
    shortages 12; sources 130, 197;
    systems 112–13
waterworks 22
weak and strong sustainability 55–6, 59
wealth 13
Wegener, M. 120–1
welfare 55, 109, 198, 203–6, f.122
well-being 181, 186
Western Bypass 116
Wijkman, A. 184
Wijkmark, B. 33–4, 41

wildlife sanctuaries 57
Wilhelmsson, M. 139
Wilkinson, C. 83
Wilkinson, R. G. 181, 183
windows 135–6
Winner, L. 190
World Bank 29
World War I 22, 24–5
World War II 24–6

Xu, J. 90

Ystad 183
Ytterfors, S. 137

Zalejska-Jonsson, A. 137–8
Zetterberg, A. 85, 87
zone-based systems 117